Power Systems

Electrical power has been the technological foundation of industrial societies for many years. Although the systems designed to provide and apply electrical energy have reached a high degree of maturity, unforeseen problems are constantly encountered, necessitating the design of more efficient and reliable systems based on novel technologies. The book series Power Systems is aimed at providing detailed, accurate and sound technical information about these new developments in electrical power engineering. It includes topics on power generation, storage and transmission as well as electrical machines. The monographs and advanced textbooks in this series address researchers, lecturers, industrial engineers and senior students in electrical engineering.

Power Systems is indexed in Scopus

Ibrahim Anwar Ibrahim · Mohammadreza Shafiee ·
Afaq Hussain · Farid Moazzen

Microgrids Design and Operation

Guiding Insights and Best Practices for Microgrids Development

Springer

Ibrahim Anwar Ibrahim
Institute of Sustainable Futures
University of Technology Sydney
Sydney, NSW, Australia

Mohammadreza Shafiee
Department of Electrical and Computer Engineering
University of New Brunswick
Fredericton, NB, Canada

Afaq Hussain
School of Electrical and Data Engineering
University of Technology Sydney
Sydney, NSW, Australia

Farid Moazzen
School of Electrical and Data Engineering
University of Technology Sydney
Sydney, NSW, Australia

ISSN 1612-1287 ISSN 1860-4676 (electronic)
Power Systems
ISBN 978-981-96-4916-7 ISBN 978-981-96-4914-3 (eBook)
https://doi.org/10.1007/978-981-96-4914-3

This Springer imprint is published by the registered company Springer Nature Singapore Pte Ltd.
The registered company address is: 152 Beach Road, #21-01/04 Gateway East, Singapore 189721, Singapore

Foreword

This is an important and timely book. The clean energy transition is now well underway and, driven by the significant reduction in costs for renewable energy generation and the need for reducing the greenhouse intensity of the electricity supply system. The challenges to this transition are multiple: technical, regulatory, skill shortages, and the institutional inertia associated with an electric power system that in its basic form, had remained unchanged for over 120 years.

That is why this book matters. Microgrids, in the huge range of configurations described here, will be a crucial component of the transition. In a saying often attributed to Einstein "We cannot solve our problems with the same thinking we used to create them". In this case, we will not be able to create the new power system in the same way, with large, centralised generators despatching power outwards on transmission lines to clustered sites of demand. The future grid will be two-way, and will have ubiquitous sites of embedded generation, in the form of rooftop photovoltaics, neighbourhood or on-premises stationery batteries, and very soon, vehicle batteries connected to bidirectional chargers. Combine this with extreme energy efficiency and demand responsive, smart loads, and the need and potential for microgrids become clear. This provides the next level of optimisation of these elements and has massive potential to reduce costs and increase reliability.

Microgrid options, optimised appropriately, will enable renewable energy to be brought into the grid faster and cheaper, as it will reduce the costs and delays associated with large-scale transmission projects.

This book provides a how-to guide, a manual if you will, for practitioners and researchers who are wanting to support the rapid introduction and spread of microgrids into new applications and to extend existing use cases. It not only is solidly grounded in the power engineering but also has a strong emphasis on the market aspects, the modelling, the socio-technical, and the regulatory issues that inevitably are important to expand the field.

The authors are all experts, with strong practical experience as well as the rigour of their research background behind them. They have also canvassed widely many experts in the process of developing the book, referencing thoroughly a number of use cases from around the globe. That makes the book a "one-stop-shop" for those either wanting to get into this field and get an understanding of the state of the art, or who want to extend their knowledge and value in the field.

Ultimately, we will need to transition to a 100% renewable, reliable, and affordable power system, and all within the next decade or so. This book is a significant contribution to supporting that transition.

Prof. Stuart White
Director
Institute for Sustainable Futures
University of Technology Sydney
Sydney, NSW, Australia

Preface

The global energy landscape is undergoing a profound transformation, driven by the imperative to decarbonise, decentralise, and digitise power systems. In this context, microgrids have emerged as critical enablers of sustainable energy systems, offering a pathway to integrate distributed energy resources (DERs), enhance grid resilience, and empower local communities. This book, *Microgrids Design and Operation: Guiding Insights and Best Practices for Microgrid Development*, reflects years of dedicated research, practical application, and collaborative learning aimed at unravelling the complexities and unlocking the potential of microgrid technologies.

The genesis of this work lies in extensive literature reviews, global case studies, and innovative projects to provide a unique opportunity to examine microgrid feasibility in remote and regional contexts, focusing on their ability to provide reliable, cost-effective, and clean energy solutions. The lessons learned from these projects have informed the chapters of this book, ensuring its relevance to diverse stakeholders, from policymakers and utility operators to engineers, researchers, and community leaders.

The book spans the entire lifecycle of microgrid development, from conceptual design and techno-economic modelling to operational strategies and future research directions. It addresses the multifaceted challenges of integrating DERs into modern power systems, including optimisation of energy flow, regulatory compliance, market participation, and advanced energy management systems. Emerging technologies such as digital twins and artificial intelligence are explored in depth, showcasing their transformative potential for microgrid control, monitoring, and forecasting.

Recognising that the success of microgrids depends not only on technical solutions but also on socio-economic considerations, this book emphasises the importance of community engagement, equitable access, and the creation of sustainable business models. It presents real-world examples of microgrid deployments in urban, rural, and remote areas, illustrating the diverse applications and benefits of these systems across different geographies and use cases.

As you embark on this journey through the pages of this book, we hope it serves as a beacon of knowledge, sparking innovation and collaboration in the pursuit of sustainable energy systems. Whether you are a seasoned professional or a newcomer to the field, we trust that the insights, best practices, and future directions outlined here will inspire you to contribute to the evolution of energy systems for a better tomorrow.

Sydney, Australia Ibrahim Anwar Ibrahim

Acknowledgements This book is prepared based on an extensive literature review study conducted for the project titled "Transferrable Benefits from Feasibility Study of Derwent Bridge Microgrid". This project was funded by the Australian Government's Department of Industry, Science, Energy, and Resources through the Regional and Remote Communities Reliability Fund. The views expressed in this document do not necessarily reflect the views of Australian Government's Department of Industry, Science, Energy, and Resources, and the University of Technology Sydney.

The authors would like to thank A/Prof. Kaveh Khalilpour, Prof. Jahangir Hossain, and A/Prof. Li Li for their thorough review of the book's content. Additionally, the authors would like to acknowledge Dr. Yunqi Wang, Dr. Melissa Mikaela Eklund, Dr. Nabil Mohammed, Dr. Luay Dwaikat, and Dr. Felipe Monteiro for their insightful remarks and valuable feedback on this book.

Competing Interests The authors have no competing interests to declare that are relevant to the content of this manuscript.

About This Book

The global shift towards sustainable energy systems has catalysed the development and integration of microgrids, which offer resilient, flexible, and efficient solutions for modern energy challenges. *Microgrids Design and Operation: Guiding Insights and Best Practices for Microgrid Development* is a comprehensive resource that encapsulates the latest advancements, practical insights, and best practices in the field of microgrids. Designed for professionals, researchers, policymakers, and students, this book serves as both an educational guide and a practical reference for anyone interested in microgrid development and operation.

This book is structured to provide a holistic view of microgrid systems, covering their design, operation, and optimisation. It begins with foundational concepts, including definitions, types, and operation modes of microgrids, setting the stage for more advanced topics. Subsequent chapters delve into real-world applications, business models, and the integration of distributed energy resources (DERs) to ensure efficient energy management and grid resilience. Case studies from diverse geographic and operational contexts enrich the content, offering readers a tangible understanding of microgrid implementation challenges and solutions.

Key features of the book include:

- **Comprehensive Coverage**: Detailed exploration of technical, economic, and social aspects of microgrids, including advanced modelling techniques, economic analysis, and market participation strategies.
- **Emerging Technologies**: Insights into cutting-edge technologies such as digital twins, artificial intelligence, and machine learning applications in microgrid management.
- **Real-World Applications**: Practical use cases and lessons learned from national and international microgrid projects, providing a valuable perspective on global best practices.
- **Research Roadmap**: Identification of key challenges, research priorities, and future opportunities to advance the development and deployment of microgrids.

This book is the result of extensive research, drawing on contributions from leading experts in the field and informed by global projects. It reflects the collective experience and expertise of its authors in tackling the technical and socio-economic dimensions of microgrids.

Whether you are an industry professional seeking to design or optimise microgrid performance, a policymaker aiming to support sustainable energy systems, or a student exploring the future of energy, this book provides the knowledge and tools to make a meaningful impact. By bridging theory and practice, it empowers readers to navigate the complexities of modern energy systems and contribute to a more sustainable, decentralised, and resilient energy future.

Contents

About the Authors

Ibrahim Anwar Ibrahim received the B.Sc. degree in Electrical Engineering from An-Najah National University, Nablus, Palestine, in 2014, the M.Sc. degree in Electrical, Electronic and Systems Engineering from Universiti Kebangsaan Malaysia, Bangi, Malaysia, in 2016, and the Ph.D. degree in Electrical Engineering from Macquarie University, Sydney, Australia, in 2022. Ibrahim is a Senior Research Consultant within the Energy Futures team in Institute for Sustainable Futures (ISF), University of Technology Sydney (UTS). Prior to joining ISF, Ibrahim was a Research Associate with the School of Electrical and Data Engineering, UTS, and a Visiting Researcher with CSIRO Energy Centre. His main research interests mainly lie in the scope of renewable power systems, microgrids, smart grids, consumer/distributed energy resources (C/DER) systems modelling and optimisation, energy economics, and artificial intelligence (AI) and machine learning applications in Energy. Ibrahim is a member of the Institute of Electrical and Electronics Engineers (IEEE), the Institution of Engineering and Technology (IET), the Australian Institute of Energy (AIE), the Electric Energy Society of Australia (EESA), the Engineers Without Borders (EWB) Australia, the Australian Meteorological and Oceanographic Society (AMOS) and the Jordanian Engineers Association.

Mohammadreza Shafiee holds a B.Sc. in Power Systems, an M.Sc. in Control Systems, and an M.Sc. in Electrical Engineering from the University of New Brunswick, Canada, where his research focused on enhancing microgrids resilience. With over 13 years in the electrical engineering field, Mohammadreza has contributed to distribution and transmission grid optimisation, energy management systems, and the integration of renewable resources. His expertise spans developing energy models, performing economic analyses, and conducting comprehensive power system studies. His research interests include microgrids, smart grids, photovoltaic (PV) systems modelling and optimisation, power quality, and artificial intelligence for power system operation. He is a member of the Institute of Electrical and Electronics Engineers and the Emera and NB Power Research Centre for Smart Grid Technologies.

Afaq Hussain received the M.Sc. degree in Electronic Engineering from Ghulam Ishaq Khan Institute of Engineering Sciences and Technology (GIKI), Topi, Pakistan, and the B.Sc. degree in Electrical Engineering from the National University of Computer and Emerging Sciences (FAST-NUCES), Islamabad, Pakistan, in 2020 and 2017, respectively. He is currently pursuing the Ph.D. degree in Electrical Engineering with the University of Technology Sydney (UTS), Sydney, Australia. His current research interests include efficient control of power electronic converter and renewable energy generation.

Farid Moazzen holds B.Sc. and M.Sc. degrees in Electrical Power System Engineering from Iran University of Science and Technology and Tarbiat Modares University, respectively, awarded in 2012 and 2015. With diverse international experiences in the electric power industry, Farid has conducted electric load analysis, power plant connection studies, and feasibility assessments for microgrid projects, alongside his involvement in demand-side management projects and the international power and energy standards survey. He has contributed significantly to high-impact peer-reviewed journals and is currently pursuing his Ph.D. in Electric Power System Engineering at UTS. Farid's research interests encompass multi-carrier energy management, machine-learning-based renewable and demand forecasting, demand response, and electricity markets, reflecting his commitment to advancing knowledge and innovation in the field.

Abbreviations

AC	Alternating current
AEMC	Australian energy market commission
AEMO	Australian energy market operator
AER	Advanced energy resource
AI	Artificial intelligence
AMI	Advanced metering infrastructure
ANN	Artificial neural network
API	Application programming interface
AR	Augmented reality
ATLCC	Annual total life cycle cost
AVR	Automatic voltage regulator
BESS	Battery energy storage system
BESS	Battery energy storage systems
BFM	Branch flow model
Bi-LSTM	Bi-directional long short-term memory
BIM	Bus injection model
C	Carbon
CanREL	Canadian renewable energy laboratory
CCGT	Combined cycle gas turbine
CEMS	Customer energy management system
CESS	Capacitive energy storage system
CIGRE	International council on large electric systems
CIS	Customer information system
CNNs	Convolutional neural network
CO_2	Carbon dioxide
CO_2e	Carbon dioxide equivalent
CPT	Cloud prediction technology
CPUC	California Public Utilities Commission
CSIRO	Commonwealth Scientific and Industrial Research Organisation
DBOOM	Design, build, own, operate, and maintain

DBSCAN	Density-based spatial clustering of application with noise clustering
DC	Direct current
DER	Distributed energy resource
DERMS	Distributed Energy Resources Management System
DFIG	Doubly-fed induction generator
DMS	Distribution management system
DNSP	Distribution network service provider
DOD	Depth of discharge
DPSP	Deficiency of power supply probability
DRP	Demand response program
DSO	Distribution network operators
DSSE	Distribution system state estimation
DSTATCOM	Distribution system with distribution static compensator
DSVC	Distributed static var compensation
EA	Evolutionary algorithm
EDRP	Emergency demand response program
EENS	Expected energy not supplied
EESG	Electrically excited synchronous generator
EFR	Energy fluctuation rate
ELF	Equivalent loss factor
EMS	Energy management system
EMT	Electromagnetic transient
EP	Evolutionary programming
EPRI	Electric Power Research Institute
EPS	Electric Power System
ESS	Energy storage system
EVs	Electric vehicles
EVSE	Electric vehicles supply equipment
FESS	Flywheel energy storage system
FiT	Feed-in tariff
FPGA	Field programmable gate array
FSIG	Fixed-speed induction generator
FTE	Full-time equivalent
GA	Genetic algorithms
GANs	Generative adversarial network
GE	General Electric
GHG	Greenhouse gas
GIS	Geographic information system
GRUs	Gated recurrent unit
GUI	Graphical user interface
H_2	Hydrogen
HAN	Home area network
HECO	Hawaiian Electric Company
HFC	Hydrogen fuel cell

HIL	Hardware-in-the-loop
HILP	High-impact-low-probability
HMI	Human-machine interface
HOMER	Hybrid optimisation model for multiple energy resources
HTB	Hardware testbed
HV	High voltage
IEC	International electrotechnical commission
IED	Intelligent electronic device
IEEE	Institute of electrical and electronics engineers
IES	Inverter energy system
IGBT	Insulated gate bipolar transistor
iHOGA	I-hybrid optimization by genetic algorithm
IoT	Internet of things
ISO	International Organisation for Standardisation
IT	Information technologies
k-NN	K-nearest neighbour
KPI	Key performance indicator
LCOE	Levelised cost of energy
LEP	Loss of energy probability
LLP	Loss of load probability
LLSP	Loss of load supply probability
LOLE	Loss of load expected
LOLH	Loss of load hours
LOLR	Loss of load risk
LP	Linear programming
LSTM	Long short-term memory
LV	Low voltage
MAE	Mean absolute error
MAIFIe	Momentary average interruption event frequency index
MAS	Multi-agent systems
MCB	Miniature circuit breakers
MCCB	Moulded case circuit breaker
MDM	Meter data management
MDT	Microgrid design toolkit
MILP	Mixed-integer linear programming
MINLP	Mixed-integer non-linear programming
MIP	Mixed integer programming
MPC	Model predictive control
MPPT	Maximum power point tracking
MSE	Mean squared error
MTBF	Mean time between failures
MV	Medium voltage
NDE	Net dump energy
NERC	North American Electric Reliability Corporation
NLP	Non-linear programming

NM	National electricity market
NMI	National metering identifier
NPV	Net present value
NREL	National Renewable Energy Laboratory
O&M	Operation and maintenance
O_2	Oxygen
OLTC	On-load tap-changer
OMS	Outage management system
OPEN	Open platform for energy networks
OpenDSS	Open distribution system simulator
OPF	Optimal power flow
OTS	Operator training simulator
P2P	Power-to-power
PAS	Process automation system
PCC	Point of common coupling
PEMFC	Proton-exchange membrane fuel cell
PHSP	Percentage of health state probability
PMSG	Permanent magnet synchronous generator
PQ	Active and reactive control
PSCAD	Power systems computer-aided design
PSO	Particle swarm optimisation
PV	Photovoltaic
PWM	Pulse width modulation
QP	Quadratic programming
RAE	Relative absolute error
REF	Research excellence framework
RMS	Root mean square
RMSE	Root mean squared error
RNNs	Recurrent neural network
ROI	Return on Investment
RSE	Root squared error
RSP	Risk state probability
RTAC	Real-time automation controller
RTUs	Remote terminal units
RUL	Remaining useful life
SA	Simulated algorithm
SAIDI	System average interruption duration index
SAIFI	System average interruption frequency index
SAM	System advisor model
SCADA	Supervisory control and data acquisition
SCC21	Standards Coordinating Committee 21
SCGT	Single cycle gas turbine
SCR	Silicon-controlled rectifier
SGIRM	Smart grid interoperability reference model
SO	Solid oxide

SOC	State of charge
SoH	State-of-health
SPD	Surge protection devices
SPL	System performance level
SPS	Stand-alone power systems
SVM	Support vector machine
TBD	Total blackout duration
TEL	Total energy loss
ToU	Time of use
TSO	Transmission system operators
UC	Unit commitment
UL	Unmet load
UPS	Uninterruptible power supply
V2G	Vehicle-to-grid
VFD	Variable frequency drive
VPN	Virtual private network
VPP	Virtual Power Plant
VR	Virtual reality
VSI	Voltage source inverters
WW	Wattwatchers

List of Figures

List of Tables

Chapter 1
Introduction

1.1 Background

Electrification and other factors will be driving the enormous growth in global electricity consumption over the next coming decades. It is the main motive to move towards a decarbonised, decentralised and digitalised energy systems [1]. The most suitable location to build a microgrid is where the reliability of power supply from a conventional grid is comparatively lower, there are higher power losses and the cost of supply of energy is higher. In recent years, microgrids are coming online to connect on-site distributed energy resources (DERs) to local facilities. Many pilot research and development projects, combined with standards development activities, have been completed over the past decade to broaden the knowledge as well as the application of microgrids. These projects deal with various practical applications of microgrids for resiliency, mostly with islanding capabilities during outages, at a range of locations such as universities, military bases, communities, and critical infrastructure sites. These activities have led to a better understanding of microgrids as planned entities with unique functionality for control and operations, and, more importantly, with unique requirements in their interconnection and interaction with the grid, allowing active deployment to address issues and needs associated with their power supply and demand [2].

Australian Energy Market Operator (AEMO) forecasts that 13–22% of total underlying annual national electricity market (NEM) consumption could be provided from DERs by 2040 [3]. In addition, AEMO states that further investment is needed in large-scale DERs, 26–50 GW, to replace the retiring conventional plants by 2040. It is also forecasted that distributed and grid-scale photovoltaic (PV) systems will have around 61% contribution to the total new renewable energy capacity as of 2040. Moreover, the amount of capacity that is scheduled and dispatchable will be less than 50% by 2040, unless some form of orchestration is implemented in small-scale assets [4]. Therefore, technical, regulatory, and economic investigations, led by AEMO, are being carried out to maximise the installations of DERs to allow end-users benefit

I. A. Ibrahim et al., *Microgrids Design and Operation*, Power Systems,
https://doi.org/10.1007/978-981-96-4914-3_1

from their investments and to maximise the DERs' contribution to the reliability, security, and resilience of the power system.

Microgrids are essential components of the evolution from centralised bulk power systems to decentralised systems that can integrate DERs for energy and capacity as well as for flexibility and grid services. Today, microgrids can interchange energy with distribution networks across the point of common coupling (PCC), provide flexibility to grid services, and participate in the market. Further, microgrids can offer several services to their overlay grid, thereby establishing new markets and improving overall energy efficiency [5].

In addition, microgrids are great ways to integrate small-scale DERs on the low-voltage distribution networks and provide customers with the flexibility to contribute to electricity generation. The local control of the infrastructure encourages increased investment from private sectors and creates microgrid hotbeds for entrepreneurship [6]. In addition, it is a feasible and viable option for supplying power to remote and regional communities where power transmission over long transmission lines from centralised generation is extremely expensive or even impossible. Microgrids can deliver ancillary services, for example, peak shaving, voltage control, and black start to electrical grids during its grid-connected mode either individually or in clusters of microgrids called multi-microgrids [7]. During its islanded mode, it can supply power to critical loads from local generation and thereby contribute to improving the resilience of the power supply.

1.2 Overview

A microgrid is defined as a group of DERs and/or energy storage systems (ESSs), servicing controllable and uncontrolled loads, and integrated locally as a single controllable entity that include controls and communication links [2, 8, 9]. AEMO defines microgrid [4], as:

> … a small-scale power system that consists of distributed generation sources that are [sic] linked to an intelligent communication and control system to supply power to distributed loads. They are usually operated autonomously to be part of the main electricity network or switched to be 'islanded' depending on their type and operational scenarios …

The difference between a passive grid and a microgrid supplied by distributed small-size sources is rooted in the principle of coordination and management techniques of accessible assets [6]. Multiple loads and generation sources, potentially owned or operated by several parties, also makes microgrids distinct from stand-alone power systems (SPSs). Microgrids can connect to the grid at a PCC and maintain frequency and voltage at the same level as the main grid [4]. Contrary to virtual power plants (VPPs), the main purpose of microgrids is to balance supply and demand. A microgrid always includes a distribution network that needs to be controlled. Another key difference is its ability to operate in islanding mode and autonomously switch from grid-connected to islanded operation and vice versa. To operate in islanded mode, the microgrid needs a well-planned mix of storage devices, intelligent loads, and generation, to establish a balance between supply and demand [6].

Microgrids can be operated in both grid-connected and islanded modes [10, 11]. In grid-connected mode, microgrids have a PCC with a host grid, which is typically a distribution system, and a seamless transition to islanded operation should be possible. Islanding mode can occur to grid-connected microgrids either intentionally (e.g., for maintenance or operating/cost reasons) or unintentionally (due to faults or system failures). Therefore, a microgrid must satisfy certain standards of reliability, power quality, and adequacy in both grid-connected and islanded operation modes [2].

The microgrid can be designed and operated in multi-feeders using intelligent switches and high-speed communication links compared with a single feeder in the conventional microgrid. The purpose of multi-feeders is that if there is a fault in the primary feeder, the microgrid can connect to the available non-faulty feeder to increase the efficiency and reduce the microgrid downtime [5]. The outer boundary of the microgrid is dynamic and can be adjusted based on the data that the central controller receives from the smart switches. This option is more reliable than the conventional solutions. However, it is costly and strongly depends on the communication links. Thereby, if there is any communication failure, the operation of such option can be affected negatively. Therefore, backup channels should be considered to avoid any communication failures, such as distribution control strategies.

To sum up, a microgrid can be defined based on the following main characteristics [2]:

- A "small" grid from a few kW to several MW.
- A local grid serves well-defined, contained areas, such as houses, factories, institutions, or urban/remote communities.
- A local grid serving an identified region with enough capacity to supply all or the majority of the local grid's loads (e.g., critical loads).
- It operates at a distribution system level (i.e., medium voltage and/or low voltage) with possible imbalances due to unbalanced loads.
- It contains various types of DERs.
- It can operate independently from the grid (i.e., islanded operation mode).
- It is controlled or coordinated by a microgrid central control and communicates through links and protocols, which makes it a good example of cyber-physical systems.
- It may contain voltage regulators, such as the on-load tap-changer (OLTC), voltage compensating transformers, or capacitor banks that are used in distribution feeders; however, the availability of DERs with voltage regulation capabilities and very short lines in most microgrids renders them obsolete.
- The dispatchable DERs (e.g., diesel generators, hydropower, battery energy storage systems (BESS), etc.) in the microgrid should have voltage and frequency regulation capabilities to operate the microgrid in islanded mode [12].

- As a result of the development of a "plug-and-play" interconnected power grid, microgrids are expected to facilitate the installation or deployment of future renewable energy systems, as well as allow efficient control and optimum benefits from the power generated by DERs [13].

1.3 Scope

Microgrids have received considerable attention in the past two decades due to the increased use of inverter-based resources, including renewable DERs and ESSs, driven by environmental issues, providing more reliably and resilience services and secure green energy access in remote communities. Accordingly, several research and pilot projects nationally and internationally have been completed with considering the standards development activities in the past 10 years. These projects are conducted for several practical applications, especially to increase the reliability of the service and the resiliency during outages in different locations, including rural areas, universities, and military bases. These projects have led to better understating of the designing and operation plans, provided several lessons learnt and recommendations as well as listed the expected risks and challenges. Thus, the unique functionalities and features of microgrids in controlling and operating their assists with unique requirements in the interconnection and interaction with the main grid leading to address several issues that the current microgrids are facing in balancing the demand-side and supply-side. Furthermore, the current activities of microgrids in exchanging the energy with the main grid, providing ancillary services, and participating in the energy market will lead to significant advancement toward fully integrated with the main grid as an active participant in the power delivery system. Therefore, consistent framework is needed around the country for designing and operating the microgrids to allow the peer learning and experience exchange in order to overcome the national and international challenges and limitations. Based on that, a capability review for 20 national and 10 international microgrid projects as well as several technical reports published by several national and international research institutes and departments (e.g., AEMO, Commonwealth Scientific and Industrial Research Organisation (CSIRO), Electric Power Research Institute (EPRI), National Renewable Energy Laboratory (NREL), International Council on Large Electric Systems (CIGRE), United States Department of Energy and etc.) are conducted to collect the lessons learnt and best practices to prove a research road map in order to extend the understating of microgrids as shown in Fig. 1.1.

Accordingly, the national and international projects are reviewed to collect the lessons learnt, challenges and best practices in modelling, analysing, optimising, controlling, and managing the energy in the real-time domain with the implementation of data acquisition and digital twin technology considering the standards and requirements to achieve the use case(s). In addition, the implemented business models and energy market structure for each project are reviewed to extend the understanding of the participation in the energy market and the engagement of the main actors. These

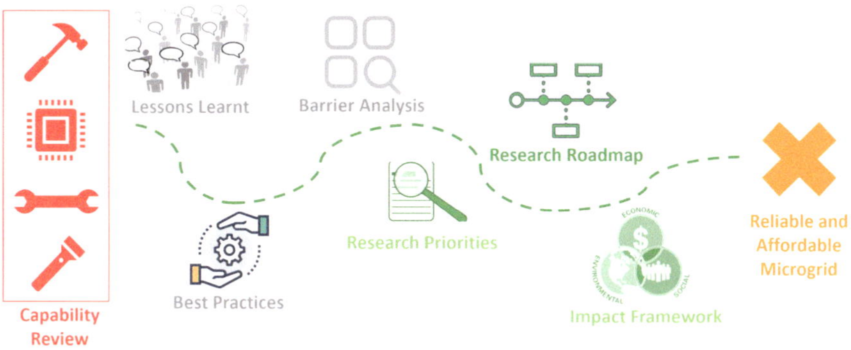

Fig. 1.1 High-level review scope

national and international projects are summarised in Appendix A and Appendix B, respectively.

Accordingly, several barriers are defined and associated with the research priorities to overcome these barriers. Based on that, a research roadmap has been set with defining several milestones and their research activities. Finally, a framework has been defined to assess the impacts of the microgrid on the adoption of new technology, skills development, technical, economic, and social aspects to achieve a reliable and affordable microgrid.

1.4 Types and Operation Modes of Microgrids

Microgrids can be categorised based on different criteria, such as geographical location, application, scale, architectures, types of current flow and operation. In terms of the geographical location and/or application, the microgrids can be classified into urban/utility microgrids (e.g., community, commercial/industrial and campus/institutional), rural microgrids, and military microgrids [14]. The urban/utility microgrids are usually designed to work in grid-connected mode and are capable to operate in an islanded mode for short periods of time (e.g., an outage in the distribution feeder). Although rural microgrids are connected to the main distribution networks, they are differing from urban/utility microgrids in the characterisation, which they mostly characterised by a long feeder that feed populated rural areas. Usually, rural microgrids have significant voltage fluctuations and unbalanced. Another type is the military microgrids, which are designed to provide power to military applications with sensitive loads. Therefore, military microgrids require complex configurations to ensure high reliability.

Moreover, microgrids can also be classified based on the scale of DERs and loads that they manage. Accordingly, a microgrid can be referred to as “mini-grid”, “nano-grid” or “pico-grid”. This size can include an aggregation of controllable loads in a dwelling (i.e., “pico-microgrid”), managing DERs and loads in a building (“nano-grid”) or managing DERs and loads in a farm (e.g., “mini-grid”). In terms of architectures, microgrids can be categorised into three main options, such as radial, ring (i.e., loop), and mesh. Each architecture has its own advantages and disadvantages based on the costs, operation, installation, and maintenance. Radial microgrids are the most popular as they are simple and can suit several applications, however they are not resilient to fault. Ring microgrids can overcome this issue by offering two connections to the distribution network. As some applications (e.g., military microgrids) require high reliability, mesh microgrids can be the best solution as they have three connections or more to the distribution network. But they are more expensive and complex than the radial and ring microgrids. The classification of the microgrids in terms of types of current flow and operation can be considered as the popular classification criteria in the literature. Therefore, the different categories of microgrids based on both criteria are explained in detail in this section.

As success criteria, the microgrid under any category must be able to: (i) manage the demand and supply sides under normal operating conditions; (ii) operate in an islanded mode either under emergency, maintenance and/or fault conditions (i.e., forced by the main grid), or as a local decision to secure some financial benefits; (iii) provide the energy to the critical/sensitive loads and shed non-critical/non-sensitive loads based on a priority list; and (iv) reconnect to the main grid once the upstream fault has been cleared [13]. In general, microgrids operate in both grid-connected and islanded modes. In grid-connected operation mode, the microgrids have a point of interconnection, typically referred to as PCC, with a host grid, which is typically a transmission and/or distribution system and should be capable of a seamless transition to islanded operation mode following the requirements of the transition mode [2].

1.4.1 Types of Microgrid

In general, microgrids are classified based on the types of current flow (e.g., alternating current (AC), direct current (DC), or hybrid) and bus connections between the DERs, ESS and loads in the microgrid.

1.4.1.1 AC Microgrid

AC microgrids represent the AC source in the microgrid and can be easily connected to the utility grid via AC bus without the need to use bi-directional power electronic converters. The importance of AC microgrids in the literature has increased in recent years due to their compatibility with already established AC distribution grids. In most cases, AC microgrid adopts the voltage and frequency according to the utility grid or consumer load demand within the microgrid [15]. The disadvantage of the AC microgrid is that it is difficult to operate and control under various conditions compared with DC microgrids. AC microgrids can be classified into three categories based on the type of connection with the distribution systems. These categories are single-phase, three-phase without neutral, and three-phase with neutral [16].

1.4.1.2 DC Microgrid

The DC microgrid represents the DC power source in the microgrid and has emerged in the literature due to its numerous applications in modern electronic devices and the accessibility of DC renewable energy sources, such as solar and fuel cells. The DC bus is connected to the main grid through a bi-directional DC-AC converter for power exchange for maintaining the power balance. The advantage of a DC microgrid compared to an AC microgrid is its higher transmission and conversion efficiency and better short circuit protection [17]. DC microgrids are common and commercialised in telecommunications systems, naval vessels, and electric aircraft [18].

1.4.1.3 Hybrid Microgrid

A hybrid microgrid represents both AC and DC sources in the microgrid. The AC bus can be connected directly to the main grid without the need for bi-directional power converters. While the DC bus integrates the DC DERs, EES and electric vehicles supply equipment (EVSE), and supplies the power directly to the DC loads if any existed. The advantages of the hybrid microgrid are to increase reliability, minimise conversion steps and reduce system costs [15]. The overall efficiency of the hybrid microgrid is increased by the benefits defined above. The consumers within the hybrid microgrid have the options of both an AC and a DC source and they can use either source according to their needs [19]. Hence, hybrid microgrid has both advantages of AC and DC microgrids, which overcome their both disadvantages.

1.4.2 Operation Modes

The microgrid can mainly operate in two modes: islanded and/or grid-connected operation modes. The microgrid can also transfer between both modes using connection/disconnection switches and controls, which can be referred to as transient mode. Here, the islanded and grid-connected operation modes are defined with considering the transient mode requirements.

1.4.2.1 Islanded Mode

Local generation resources combined with storage devices that operate in grid-forming mode allow microgrids to operate in islanded mode. The islanded operation mode can take place intentionally if maintenance occurs in the upstream network, or unintentionally if there is a disruption in the upstream network, which results in the microgrid operating autonomously and feeding part or all its loads based on a priority list. The main task of islanding is to provide power to the critical/sensitive loads. If excess power is available in the microgrid, the microgrid can then expand its outer boundary based on the available power to provide power to non-critical/non-sensitive loads. Dispatchable DERs (e.g., diesel generators, hydropower, BESS, etc.) play an important role in power balancing by responding fast to initial load unbalance after islanding, and adjusting to load changes generation [2]. Active monitoring and boundary management control are essential in this mode to ensure power balance and acceptable voltage/frequency deviation [5]. For instance, the microgrid controller comprehensively monitors the state of charge (SOC) of the battery and when the SOC is close to the lower limit, the controller stops the power flow from the battery to the microgrid and in this state; the diesel generator, as an example of dispatchable DERs, is responsible for voltage regulation. Similarly, when the SOC is close to the upper limit, the controller sends a command to shut down the diesel generator.

1.4.2.2 Grid-Connected Mode

Grid-connected operation mode stands for connecting the microgrids to any utility feeder. For the integration of microgrids into the utility power grid, voltage, frequency, and phase angle should be within the predefined limits on both sides. During grid-connected mode, the microgrid extends its limits to power both critical/sensitive and non-critical/non-sensitive loads [5]. The BESS supplies power to microgrids even in grid-connected mode and helps with energy management of the power system (e.g., peak load control).

In normal grid-connected operation, the microgrid is capable of economic dispatch to optimise the utilisation of renewable power for local-served loads. In case of the microgrid connects to the grid through multiple PCCs (e.g., one feeder serving as a primary PCC and two backups) and the electricity has similar or even lower prices than renewable energy, the backup feeders will serve their own loads, while the main feeder balances power deficits between power generation from the DER and loads. If renewable energy has a lower price than electricity and excess renewable energy is available, the next logic could be followed [5]:

- The microgrid can extend its boundary and utilise renewable power to serve more loads than the base case.
- This operation is realised through multi-objective optimisation algorithm that can be implemented in the microgrid central control.
- Microgrid central control collects local information from each microgrid asset and sends back commands to achieve power balance and microgrid reconfiguration.
- The microgrid central control also identifies the optimal boundary and determines the switching commands according to the available DERs generation, available capacity and SOC of the battery, and size of dispatchable load chunks.

1.4.2.3 Transition Mode

Switching the microgrid operation from grid-connected mode to islanded mode may cause severe issues as the balance between supply and demand-side may not be met if the internal power generation in the microgrid is not sufficient. In addition, switching from islanded mode back to grid-connected mode requires maintaining the reference voltage and frequency of the main grid to be able for synchronisation. Accordingly, several grid codes/requirements should be considered to be able to avoid any longer-term issue inside the energy provision. These requirements are:

Post-fault operation

The microgrid can quickly isolate a fault and reconnect to another optimally selected healthy utility feeder by the smart switches and/or multiple feeders. Smart switches are equipped with sophisticated protection algorithms, which are capable of autonomous fault isolation and grid reconnection. If the fault is located within the microgrid boundary, three steps are likely to be involved [5]:

- The fault is detected, located, and isolated by opening smart switches.
- The microgrid disconnects from the grid and goes to islanded mode, where the BESS or backup generator regulates the AC voltage and balances power generation and loads within the microgrid.
- The BESS or backup generator tracks the voltage of the selected healthy utility feeder. When the magnitude, frequency, and angle differences between the microgrid and grid voltages drop into a desired range, microgrid reconnects back to the main grid.

If the fault is located outside of the microgrid, the operating mode of the microgrid will not be affected, and the unsupported loads of the faulted grid will be quickly picked up by closing relevant switch if sufficient power generation is available.

Black start

Black start capability involves powering the microgrid correctly and safely after a grid outage, without any external feeder power. It also refers to the ability of an inverter unit(s)/farm(s) to start with the absence of external voltage reference (e.g., grid voltage). Active and reactive powers balance is critical during the black start to ensure stable, balanced power between demand and supply sides within the microgrid and guarantee that voltage and frequency stay inside the normal operating range while shortening the time it takes for the process [5]. In order to minimise the loss of power supplied in a black start procedure, a sequence of actions must be identified and followed. This must be done by following the safety guidelines/requirements for the utility crew and equipment. In a microgrid, a set of black start rules can be built into the management levels as follows [2]:

- All network loads should be disconnected to minimise the magnitude and unpredictability of frequency deviations as well as voltage transients during the service restoration phase.

- Create islands by segmenting the distribution network around each microgrid and generator with black start capability.
- Energising the network gradually to avoid large variations in frequency and voltage.
- Synchronisation of various islands by energising the branches between the energised islands under synchronism conditions.
- Connection of priority loads (e.g., critical/sensitive loads).
- Microgrid synchronisation with the MV network.
- The remaining MV/LV transformers will be gradually energised, which will allow the remaining loads to be gradually reintegrated.
- Reconnection of the load after the entire network has been energised by increasing power from dispatchable DERs.
- Non-dispatchable DERs can be connected to the MV network when it is powerful enough to handle the power fluctuations from variable non-dispatchable DERs.
- Finally, link the MV microgrid to the upstream HV grid, assuming the HV grid is ready for synchronisation.

Grid re-synchronisation and re-connection

Once the supervisory control and data acquisition (SCADA) system or any other monitoring and controlling platform confirms the grid's readiness, the microgrid central control will set the voltage reference of the local controllers to control the inverters or available backup generators to synchronise with the grid voltage. With certain active re-synchronisation schemes, the local controllers need to rely on the instantaneous angle and voltage amplitude signals from the main grid, requiring high-speed communication [5].

Summary

AC microgrid is preferable giving the fact that most of the loads are operated in AC domain. However, hybrid microgrids are showing promising solutions to overcome the limitations of AC and DC microgrids. In this context, the grid-connected mode is considered as the most economic operation mode, as it reduces the investment and operation costs of the dispatchable DERs and ESSs. Nevertheless, the microgrid should be designed to be able to operate in grid-connected and islanded mode to handle any abnormal event that could be occurred, which requires either forced or optional disconnection from the main grid. In islanded mode, the microgrid should be able to operate for a specific time horizon, which can be defined based on the application, use case and budget. In addition, the microgrid design should consider post-fault operation, black start, grid re-synchronisation and re-connection requirements. Accordingly, the microgrid voltage and frequency should be maintained by the microgrid local DER units and/or ESS units.

References

1. Schneider Electric (2018) Digital grid unleashed. Andover, MA, USA
2. Sub-Committee on MicroGrids Whitepaper, IEEE Smart Grid Technical Activities Committee (2022) Microgrids: utility challenges and opportunities
3. AEMO (2020) Integrated system plan for the national electricity market
4. Economics and Industry Standing Committee (2019) Implications of a distribution energy future. Western Australia, Australia
5. EPRI (2018) A smart and flexible microgrid with dynamic boundary and intelligent open-source controller. Palo Alto, California, USA
6. Eger K, Götz J, Sauerwein R et al (2011) Microgrid scenario building blocks. Munich, Germany
7. Lopes JAP, Moreira CL, Madureira AG (2006) Defining control strategies for MicroGrids islanded operation. IEEE Trans Power Syst 21:916–924. https://doi.org/10.1109/TPWRS.2006.873018
8. Hatziargyriou N, Asano H, Iravani R, Marnay C (2007) Microgrids. IEEE Power Energy Mag 5:78–94. https://doi.org/10.1109/MPAE.2007.376583
9. Lasseter B (2001) Microgrids [distributed power generation]. In: 2001 IEEE power engineering society winter meeting. Conference proceedings (Cat. No. 01CH37194). IEEE, pp 146–149
10. Farrokhabadi M, Canizares CA, Simpson-Porco JW et al (2020) Microgrid stability definitions, analysis, and examples. IEEE Trans Power Syst 35:13–29. https://doi.org/10.1109/TPWRS.2019.2925703
11. Olivares DE, Mehrizi-Sani A, Etemadi AH et al (2014) Trends in microgrid control. IEEE Trans Smart Grid 5:1905–1919. https://doi.org/10.1109/TSG.2013.2295514
12. IEEE Power and Energy Society (2017) IEEE Standard for the specification of microgrid controllers in IEEE Std. 2030.7
13. Rahman S, Pipattanasomporn M (2010) Modeling and simulation of a distributed generation-integrated intelligent microgrid. Arlington, VA, USA
14. Australian Power Quality and Reliability Centre (2021) Microgrids: concepts & technologies. Wollomgong, NSW
15. Hossain MA, Pota HR, Hossain MJ, Blaabjerg F (2019) Evolution of microgrids with converter-interfaced generations: challenges and opportunities. Int J Electr Power Energy Syst 109:160–186. https://doi.org/10.1016/j.ijepes.2019.01.038
16. De Matos JG, E Silva FSF, Ribeiro LAS (2015) Power control in AC isolated microgrids with renewable energy sources and energy storage systems. IEEE Trans Ind Electron 62:3490–3498. https://doi.org/10.1109/TIE.2014.2367463
17. Salomonsson D, Sannino A (2007) Low-voltage DC distribution system for commercial power systems with sensitive electronic loads. IEEE Trans Power Deliv 22:1620–1627. https://doi.org/10.1109/TPWRD.2006.883024
18. Elsayed AT, Mohamed AA, Mohammed OA (2015) DC microgrids and distribution systems: an overview. Electr Power Syst Res 119:407–417. https://doi.org/10.1016/J.EPSR.2014.10.017
19. Dawoud SM, Lin X, Okba MI (2018) Hybrid renewable microgrid optimization techniques: a review. Renew Sustain Energy Rev 82:2039–2052. https://doi.org/10.1016/J.RSER.2017.08.007

Chapter 2
Key Application Use Cases in Microgrids

2.1 General

Several application use cases are collected based on the national and international practices. This section describes the most common use cases for the microgrid related to the financial and technical benefits and the provided services. Accordingly, the key application use cases are described at a high level as well as in detail associated with the main actors. The use cases are the main driver for any microgrid design, and a keystone in building the business models. Therefore, understanding the use cases for a microgrid helps to defining the design and operation plans based on the requirements and needs as well as the main actors. That provides an optimal design scenario, and business model as well as saving resources, time, and efforts. The key application used cases have been collected from the national and international projects (See Tables A1–A9 in Appendix A for national projects and Tables B1–B9 in Appendix B for international projects), including [1–9].

2.2 High-Level Use Cases

The high-level use cases are split into two main categories: "business" use cases, and "control and management" use cases. The business use cases category contains four high-level use cases, which are covering the participation in the NEM, providing several ancillary services, islanding operation mode as a paid service, and providing an open internal and external trading platform. In addition, the control, and management use cases category identifies nine high-level use cases, including the demand and supply-side management, major control tasks, black start in islanded operation, protection, and monitoring, and upgrading the infrastructure of the microgrid.

I. A. Ibrahim et al., *Microgrids Design and Operation*, Power Systems,
https://doi.org/10.1007/978-981-96-4914-3_2

Table 2.1 Overview on high-level business use cases

Use case	Description
Microgrid operator trades energy on external markets	In this use case, microgrid operator takes the role of the aggregator in representing the prosumers based on contracts to manage the financial interactions of selling and buying the energy on the external markets (e.g., wholesale market)
Microgrid provides an open trading and communication platform for both local and external markets	This use case aims to sell the energy generated from the supply-side players (e.g., DER owners) and buy the energy to the demand-side players (e.g., consumer, storage owner, etc.) using an open trading platform on the local and external markets using optimisation strategies
Microgrid operator sells balancing and ancillary services	This use case aims to sell the balancing and/or ancillary services to the main grid via the microgrid operator, such as stabilising the main grid, improving the resiliency of the main grid, etc.
Microgrid provides islanding mode	In this use case, the overlay grid operator uses the services provided by the microgrid operator to go into islanding operation mode

2.2.1 Business Use Cases

Business use cases describe the financial interactions within the wholesale market, including the roles and channels of information exchange of balancing and ancillary services between main actors and wholesale or retailer markets. Thus, a business use case can be relayed on one or more control and management use case(s) to manage the information exchange schemes. An overview on the high-level business use cases in microgrid can be listed in Table 2.1.

2.2.2 Control and Management Use Cases

Control and management use cases describe several services and activities that the microgrid can provide for balancing the supply and demand on different timescales, managing the demand, supply, and storage within the microgrid, starting the microgrid in islanding operation mode (i.e., black start). These services and activities are also including the protection, restoration, monitoring and planning to upgrade the microgrid infrastructure. The high-level control and management use cases are briefly identified and described in Table 2.2.

Table 2.2 Overview on high-level control and management use cases

Use case	Description
Balancing supply and demand on different timescales	This use case aims to monitor and control the energy flow to ensure the balancing between the supply and demand on different timescales into the microgrid to achieve a stable by maintaining the reference voltage and frequency
Demand-side management	This use case aims to monitor and encourage the consumers to change their electricity usage habits based on the energy status (i.e., availability and needs) in the microgrid using automated load management procedures
Supply-side management	This use case aims to monitor and control the DERs into the microgrid based on the activities of the utility and consumers to ensure the stability and reliability
Storage management	This use case aims to monitor and manage the charging and discharging in the storage devices in microgrids to balance volatile supply and demand
Microgrid black start in islanding mode	This use case aims to provide black start to the microgrid to operate in the islanding mode after a general system blackout utilising autonomous agent ideas based on predefined rules
Protection and restoration	This use case aims to isolate the faulted part of the power system to improve the utility reliability and power quality. In addition, it aims to restore power to the out-of-service areas as quickly as possible
Smart metering	This use aims to enable the advance metering into the microgrid to provide supply-side and demand-side functions, and several services as well as to get the customers more involved in energy management. This use case allows the microgrid control centre to monitor and control the energy flow into the microgrid in a secure and trusted manner
Auto-configuration	This use case aims to allow the new devices (e.g., DERs or intelligent appliances) or subsystems (e.g., home/building energy management systems) automatically configure themselves in the microgrid
Long-term planning of microgrid infrastructure design and upgrading	This use case aims to plan the microgrid infrastructure upgrade, including the planning tools and simulations that consider possible changes (e.g., new DERs connections, grid topology, population, and regulation)

2.3 Detailed Use Cases

The high-level business, and control and management use cases have some sub-use cases that describe the detailed benefits and services can be provided by or to the main actors in the microgrid. Here, the detailed description of the high-level business, and control and management use cases with their sub-use cases associate with the main actors are mentioned.

2.3.1 Detailed Description of Business Use Cases

Some of the business use cases have sub-use cases. Here, each high-level business use case and its related sub-use cases are described in detail associated with their main actors. An overview of the business use cases associated with their sub-use cases is shown in Fig. 2.1.

2.3.1.1 Microgrid Operator Trades Energy on External Markets

In this use case, the microgrid operator acts as an aggregator in representing all the prosumers in the microgrid to trade the energy on external markets (e.g., wholesales market). Accordingly, the microgrid operator produces the offers, bids, and negotiates with the external markets under various optimisation strategies. The sub-use cases of "microgrid operator trades energy on external markets" associated with their main actors are outlined in Table 2.3.

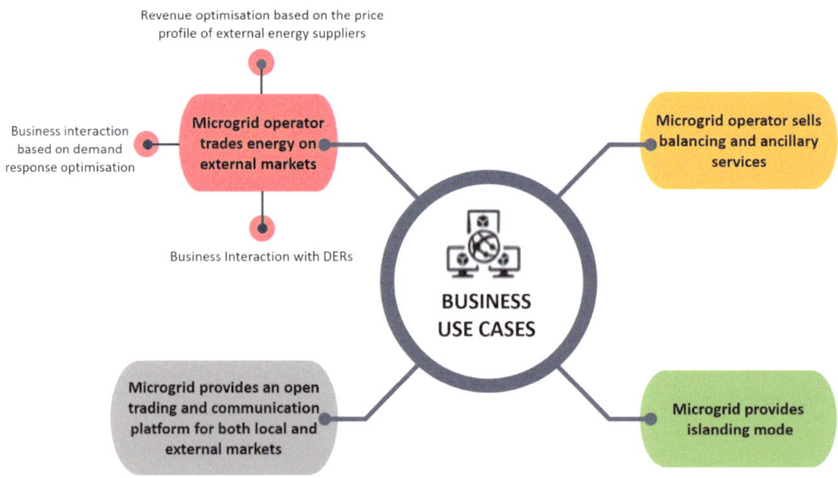

Fig. 2.1 Overview of business use cases associated with their sub-use cases

Table 2.3 Sub-use cases of microgrid operator trades energy on external markets

Sub-use case	Scope and objectives	Actors
Revenue optimisation based on the price profile of external energy suppliers	In this sub-use case, the microgrid operator optimises the revenue for aggregated customers by (i) collecting and analysing the prosumers' data; (ii) forecasting the energy generation and consumption on different timescales; (iii) forecasting the prices (i.e., costs and earnings); (iv) optimising the revenues based on the current and forecasted market prices; and (v) conducting the financial interactions between the prosumers and market	Microgrid operator, service provider, retail market provider, supply and demand aggregator, consumer, storage owner, and DER owner
Business interaction based on demand response optimisation	In this sub-use, the microgrid operator acts as an aggregator to manage the available flexibility of the consumers through their home/building energy management system and interact with the external actors (e.g., overlay grid operator). The objective is to agree on how to optimise demand response based on a combination of higher demand bid elasticity and price elasticity on the demand side in order to boost the market for microgrids, thereby improving revenue for microgrid prosumers. This sub-use case can be realised using two main alternatives: Market optimisation with short-term load shaping: Considering the external wholesale market conditions, a microgrid operator seeks to optimally match its demand to its supply Capacity reservation for managing short-term business risks: Microgrid operator keeps some reserve available (e.g., increase or decrease the demand) to use it on the time of high demand (i.e., wholesale price spikes)	Microgrid operator, overly grid operator and home/building energy management system
Business Interaction with DERs	In this sub-use case, the microgrid operator has contracts to support the main grid (i.e., import and/or export energy) whenever the overlay grid operator needs that. Therefore, microgrid operators must ensure getting the control to enable dynamic export and import limits of the DERs and storage to balance the energy flow between the prosumers and microgrid contracts	Microgrid operator, overlay grid operator, and DER owner

2.3.1.2 Microgrid Provides an Open Trading and Communication Platform for Both Local and External Markets

This use case aims to form an open trading market for both local and external markets to increase the revenues for all the key players (i.e., supply-side, and demand-side players) in the microgrid. Accordingly, the microgrid operator manages the open trade market between prosumers and the local and external markets under various optimisation strategies. In contrast to the "microgrid operator trades energy on external markets", the microgrid operator is no longer in charge of the aggregator position. This use case contributes to stable the operation of the microgrid. The main actors in this use case are microgrid operator, overlay grid operator, supply and demand aggregator, consumer, storage owner, DER owner, and transmission system operator.

2.3.1.3 Microgrid Operator Sells Balancing and Ancillary Services

This use case aims to increase the incentives for all the key players in the microgrid by supporting the main grid with the balancing, stabilising, and ancillary services. The revenues in participating in the balancing and ancillary services market are bigger than that of the retailer market. This is feasible if the microgrid operator can modulate the generation (e.g., diesel generation, hydropower generation). In addition, it is recommended to manage the regulation in the local market instead of the national market, as it is quite common for the general trend to oppose local trends in some local points.

Furthermore, this use case can also use the ancillary services, regulations, and incentives to optimise the climate-related conditions (e.g., CO_2 consumption). Here, the microgrid operator can use the incentives sachem to minimise the CO_2 consumption by encouraging the DERs owners to generate more energy and/or consumers to reduce their energy consumption. The total CO_2 budget of microgrids can be optimised using the green energy generation methods as follows:

- Switching on the fly—if possible—to green generation.
- Reducing consumption of the CO_2—extensive generation sites.

The main actors in this use case are microgrid operator, overlay grid operator, supply and demand aggregator, consumer, storage owner, DER owner, and transmission system operator.

2.3.1.4 Microgrid Provides Islanding Mode

This use case aims to switch the operation of the microgrid to islanded via the microgrid operator based on the requests from the overlay grid operator. Mostly, the request of such as a service could be an emergency in the main grid. Here, this service should be listed in a business contract between both operators. This use case is associated with control and management use case "microgrid black start in islanding mode". The microgrid operator can rely on this use case to be more independent in the operation by asking the overlay network operator to pay some money once the grid resources are not capable to supply the energy to the microgrid. The main actors in this use case are microgrid operator and overlay grid operator.

2.3.2 *Detailed Description of Control and Management Use Cases*

Some of the control and management use cases have sub-use cases. Accordingly, each high-level control and management use case and its related sub-use cases are described in detail associated with their main actors. An overview of the control and management use cases associated with their sub-use cases is shown in Fig. 2.2.

2.3.2.1 Balancing Supply and Demand on Different Timescales

In this use case, the primary duties of load balancing and power stabilisation are the foundations of a microgrid's reliable and efficient performance. In addition to a constant balance between production and consumption, voltage and frequency control are mainly responsible for electrical stability. Therefore, the microgrid control centre requires real-time measurements to monitor any electrical imbalances to control the supply and/or the demand to achieve the balancing between both on different timescales. Even in this case, microgrid operator might need to keep the load balance for some efficiency or economics. Accordingly, the microgrid operator can adjust the generation, storage, and/or consumption. Here, grid-connected operation mode can simplify this use case to keep the microgrid stable. Therefore, the microgrid operator needs to get access to the available main grid information and status reports from the demand and supply sides, including the storage energy management systems. The sub-use cases of balancing supply and demand on different timescales associated with their main actors are listed in Table 2.4.

2.3.2.2 Demand-Side Management

This use case aims to improve the microgrid reliability and stability by managing the demand-side. This use case strongly depends on the monitoring of the energy consumption as well as the optimal power flow results to operate the microgrid optimally and efficiently (i.e., real-time energy management system). Several strategies can be employed to manage the energy demand (e.g., automated load shifting, intelligent load shedding, price-based load shifting, etc.). The microgrid control centre and home/building energy management system are the main involved actors in this use case. Here, the sub-use cases of "Demand-side management" associated with their main actors are outlined in Table 2.5.

2.3.2.3 Supply-Side Management

This use case covers all the use cases related to the supply sides, including monitor and control of the decentralised DERs to ensure the optimised operation and stability

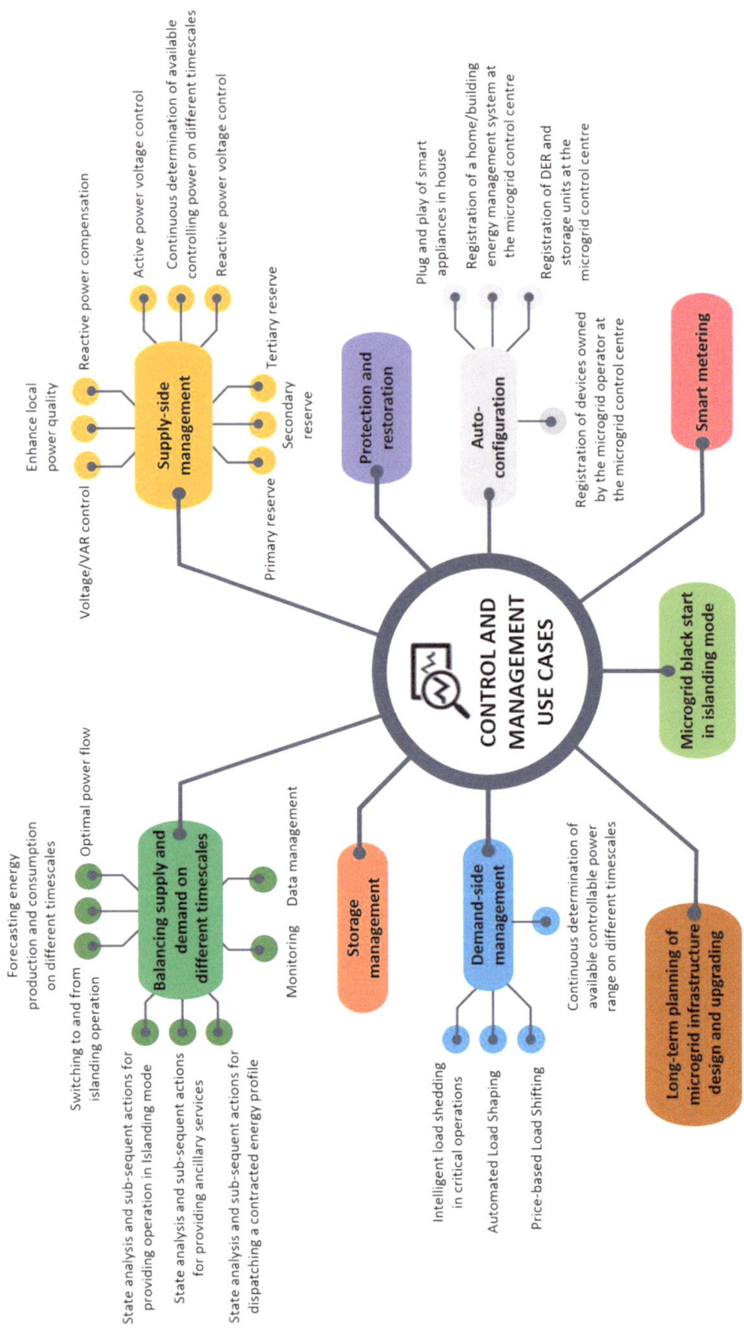

Fig. 2.2 Overview of the control and management use cases associated with their sub-use cases

Table 2.4 Sub-use cases of balancing supply and demand on different timescales

Sub-use case	Scope and objectives	Actors
Forecasting energy production and consumption on different timescales	This sub-use case aims to forecast the demand/generation profiles to allow the microgrid operator to plan for several imbalance scenarios and the associated solutions. In this sub-use case, several quantities on different time horizons can be forecasted, such as the demand and supply sides, customer level load data, weather variables, variable DERs generation and storage capacity status. This sub-use case strongly relies on the monitoring sub-use case	Microgrid operator, DER owner, consumer, storage owner, and weather forecast provider
Monitoring	This use case aims to increase the visibility of the microgrid data and information to ensure the optimal operation and future planning of the microgrid by allowing the "what-if" scenarios. The data can be measured form home/building energy management system, DERs, storage, inverters, substations, switches and regulation and pricing information on various time horizons. The measurement systems should be able to support the high sampling rate technology as some of the functions may need a high-resolution data within a short-time horizon	Microgrid control centre, home/building energy management system, DER unit, and network smart device
Data management	This sub-use case aims to manage high volumes and traffic of data (e.g., real-time measurement, historical data, forecasted data, static and dynamic operations, tariffs, etc.) as well as analysis and exchange them with the business and operational intelligence systems. Therefore, this sub-use case depends on the previous two sub-use cases to classify, clean, process, coordinate, and transfer the available data to provide useful state information for the microgrid actors. Accordingly, the control and management applications can be optimised	Microgrid control centre, home/building energy management system, DER unit, and network smart device
State analysis and sub-sequent actions for providing ancillary services	This sub-use case aims to provide the required ancillary services (e.g., active network management strategies (i.e., active power/voltage, reactive power/voltage, etc.), load shifting, control strategies (i.e., primary, secondary, and tertiary)) to stabilise the operation of the microgrid (i.e., keep the voltage and frequency within the pre-defined limits) based on the forecasted, monitored, and processed data	Microgrid control centre, DER unit, and network smart device

(continued)

Table 2.4 (continued)

Sub-use case	Scope and objectives	Actors
State analysis and sub-sequent actions for dispatching a contracted energy profile	This sub-use case aims to manage the import/export energy limits with the main grid based on the operation optimisation function(s). This process can be defined by a contract between the microgrid operator and overlay grid operator based on the business use case "Microgrid operator trades energy on external market". To utilise the import/export energy limits efficiently, the Microgrid operator should implement a demand/generation profile coordination. This coordination strongly depends on the forecasted, monitored, and managed data	Microgrid control centre, home/building energy management system, DER unit, and network smart device
State analysis and sub-sequent actions for providing operation in Islanding mode	This sub-use case aims to switch the operation of the microgrid to islanded mode either due to a fault or emergency, or to address a critical imbalance. Therefore, the microgrid control centre should be able to balance the supply and the demand within the microgrid to switch to the islanded mode with keeping the voltage and frequency within the pre-defined limits. This sub-use case can be also linked with the business use case "Microgrid provides islanding mode"	Microgrid control centre
Optimal power flow	This sub-use case aims to optimise the power flow into the microgrid on grid-connected and islanded operation modes based on single- or multi-objective functions subject to several constraints and parameters. Accordingly, the microgrid operator needs to calculate the load flow and manage the import/export limits of the DERs and storage units accordingly. This can include the flexibility of active and reactive power to achieve the objective function(s) with keeping the stability in the microgrid	Microgrid operator, DER owner, network smart device, microgrid control centre, customer (i.e., controllable load), and DER unit
Switching to and from islanding operation	This sub-use case aims to deal with switching the operation of the microgrid. By switching to islanded operation, the microgrid operator should make sure that the supply-side can meet the demand-side in the islanded operation. Otherwise, the suitable countermeasures should be internally initiated to avoid any long-term issues. By switching to grid-connected operation, the voltage and frequency should be achieved for re-synchronisation with the main grid. In contrast to "State analysis and sub-sequent actions for providing operation in islanding mode", this sub-use case allows the voltage control for synchronisation	Microgrid operator, network smart device, microgrid control centre, overly grid control centre

Table 2.5 Sub-use cases of demand-side management

Sub-use case	Scope and objectives	Actors
Continuous determination of available controllable power range on different timescales	In this sub-use case, the microgrid operator keeps monitoring any imbalance between the supply and demand through the microgrid control centre and act accordingly to balance the demand and supply sides. Either by controlling the flexible customer loads (e.g., load shifting or shaving) or by controlling the import/export limits by DER and storage units. This can also be done through the energy exchange with the main grid. Several expected scenarios should be forecasted based on the measured and historical data in the microgrid	Microgrid control centre, prosumer, and home/building energy management system
Intelligent load shedding in critical operations	This sub-use case aims to solve the imbalance issue between the demand and supply sides intelligently by following the load shedding strategy. The intelligent load shading should be able to control the controllable loads in real-time immediately according to the load significance and characteristics to balance the demand and supply sides	Microgrid control centre, customer, and home/building energy management system
Automated load shaping	This sub-use case aims to control the loads to ensure the balance between the demand and supply sides and stabilise the operation of the microgrid. This sub-use case assumes that the customers have home/building energy management systems. Such systems can control and schedule shifting automatically with offering financial incentives and environmental awareness (needs contracts with customers)	Microgrid control centre, customer, and home/building energy management system
Price-based load shifting	This sub-use case has the same aim of the previous sub-use case in order to control and schedule the shifting of the loads to ensure the stability and balancing in the microgrid. However, the response from the customers is optional based on the provided financial incentives (no contracts needed). This sub-use case is related to the high-level use cases "Demand-side management", and "Supply-side management" as well as the sub-use cases "State analysis and sub-sequent actions …"	Microgrid control centre, consumer, and home/building energy management system

in the microgrid, until the ancillary services are offered to the main grid. This contains several sub-use cases for monitoring the DERs state and controlling the DERs active and reactive power. The sub-use cases of the "Supply-side management" are related to optimising the control of import/export active and reactive powers limits within the microgrid, especially in islanded microgrid operation, or at the PCC. The main actors in this use case are microgrid control centre, DER unit, microgrid operator, and overlay grid operator. The sub-use cases of "Supply-side management" associated with their actors are outlined in Table 2.6.

2.3.2.4 Microgrid Black Start in Islanding Mode

This use case aims to allow the microgrid recovery once the microgrid could not reconnect to the main grid. Therefore, the microgrid will firstly follow a local black start strategy to recover its operation step by step. In this use case, two black start functions are included as follows:

- Local Black Start of the Microgrid after a general system blackout.
- Grid reconnection during Black Start.

The black start of the microgrid can be carried out by the cooperation of several devices based on local and centralise control strategies according to pre-defined regulations, which can be embedded in the microgrid central controller. The black start can be initiated automatically in microgrid easier than the main grid as there are not many loads, switches and DERs, which they can be controlled more flexible because of the settings power electronic interfaces. The black start should be conducted fast in real-time under extreme stress conditions. This is making the microgrid restoration fully automatic to seem quite feasible and realistic. The main actors in this use case are microgrid operator, overlay grid operator, consumer, microgrid control centre, overlay grid control centre, DER unit, storage unit, and network smart device.

2.3.2.5 Auto-configuration

This use case aims to specify and auto-configure the new devices, and sub-systems in the microgrid. These devices and sub-systems can automatically identify their capabilities and services to the microgrid control centre to be used either in providing extra services or expanding the microgrid. In this use case, the level, and rights of access the data and information of such devices as well as control them should be pre-defined based on contractual agreements to ensure the security and privacy. The main actors of this use case are microgrid operator, microgrid control centre, consumer, DER and storage owners, network smart device, and home/building energy management system. Here, the sub-use case of "Auto-configuration" use case are listed in Table 2.7.

2.3.2.6 Long-Term Planning of Microgrid Infrastructure Design and Upgrading

This use case aims to provide a planning tool, where all the data of all the components in microgrid can be located in a database. This database can also contain all the real-time measurements, forecasted data and the status of the loads, DERs, storage

Table 2.6 Sub-use cases of supply-side management

Sub-use case	Scope and objectives	Actors
Continuous determination of available controlling power on different timescales	This sub-use case aims to ensure the balancing between the demand and supply sides in the microgrid by controlling the DERs import/export limits on different timescales. Accordingly, the microgrid control centre proposes several scenarios to the microgrid operator based on the forecasted DERs generation and flexibility to control the DERs. This sub-use case relates to the sub-use cases of "Demand-side management", as the balancing between demand and supply sides can be also achieved through control the demand-side	Microgrid control centre, DER unit, microgrid operator, and overlay grid operator
Active power voltage control	This sub-use case aims to mitigate any voltage issue (i.e., under, or over-voltage issues) to keep it within pre-defined limits on all the busses in the microgrid. In this sub-use case, the microgrid control centre controls the active power import/export limits (e.g., active power curtailment). Therefore, DER active power curtailment penalties may apply. Some of distribution network service providers (DNSPs) are get the authority to do that based on the connection contract	Microgrid control centre, DER unit, microgrid operator, and overlay grid operator
Reactive power voltage control	This sub-use case also aims to mitigate any voltage issue (i.e., under, or over-voltage issues) to keep it within pre-defined limits on all the busses in the microgrid. However, in this sub-use case, the microgrid grid centre controls the reactive power import/export limits (e.g., injection or absorption of reactive power). This may curtail the active power injected to the microgrid. Therefore, DER curtailment penalties may apply	Microgrid control centre, DER unit, microgrid operator, and overlay grid operator
Reactive power compensation	This use case aims to compensate the reactive power in the microgrid to keep the power factor within per-defined limits. Therefore, the microgrid operator can control the power export/import limits of the available and flexible DERs to keep the power factor within the limits; therefore, reduce the needs for any other support devices (e.g., power factor correction unit). Also, there might be some active power curtailment. Therefore, DER curtailment penalties may apply	Microgrid control centre, DER unit, microgrid operator, and overlay grid operator
Voltage/var control	In this sub-use case, the microgrid aims to provide the voltage control services to the main grid (at the PCC) by controlling the reactive power import/export limits based on the request from the overlay grid operator. This sub-use case aims to optimise the load flow and improve the stability of the main grid. In addition, this sub-use case can also enable the dynamic operating envelopes to increase the DERs hosting capacity in the main grid	Microgrid control centre, DER unit, microgrid operator, and overlay grid operator

(continued)

Table 2.6 (continued)

Sub-use case	Scope and objectives	Actors
Enhance local power quality	In this sub-use case, the microgrid control centre aims to reduce and/or eliminate any local power quality issues within the microgrid (e.g., harmonics). Accordingly, the microgrid control centre optimises the power electronics or filters interfaced with DERs units. This optimisation can be carried out by increasing or decreasing DERs output, switching to a different operation point, adjusting the control algorithms, etc.	Microgrid control centre, DER unit, storage unit, smart device, and microgrid operator
Primary reserve	This sub-use case aims to provide primary reserve power to the main grid via the reserve power markets. Therefore, microgrid control centre, and DER and storage units should form a pool of power to offer it to the reserve power markets. The offers can be in regular basis for the main grid (tertiary requirements), while they can be very dynamic within the microgrid (primary requirements)	Microgrid control centre, DER unit, storage unit, and overlay grid operator
Secondary reserve	In this sub-use case, the microgrid provides the secondary reserve power to the main grid. The main grid considers the microgrid as a single entity (i.e., represented by the microgrid control centre); thus, the microgrid optimises the import/export limits of DERs (within 100 s) and behaviour of the flexible demand	Microgrid control centre, DER unit, and overlay grid operator
Tertiary reserve	This sub-use case aims to provide tertiary reserve power to the main grid. As in the previous sub-use case "Secondary reserve", the main grid considers the microgrid as a single entity, and microgrid optimises the import/export limits of DERs (less than 15 min) and behaviour of the flexible demand. The microgrid should be able to provide energy for at least 2 h	Microgrid control centre, DER unit, and overlay grid operator

and power electronics interface devices. This tool can be used offline to test any new scenario of the microgrid operation, or expansion of the microgrid. This tool can also be used to test any new business model, regulation and plan the future infrastructure upgrade (i.e., "what-if" scenario). The main actors in this use case are microgrid operator, microgrid control centre and forecast provider.

Summary

One of the key factors to achieving the main goals in designing and operating any microgrid is selecting the proper use case(s). Use case(s) can significantly affect the future planning in upgrading and/or expanding the microgrid. Use case(s) also can be considered as the keystone of forming the business models, legal contracts, regulations, and standards to encourage the prosumers to interact with the controlling, monitoring, and pricing signals. The use cases are categorised into "Business" and "Control and Management". The "Business" use cases are addressing the interactions

Table 2.7 Sub-use cases of auto-configuration

Sub-use case	Scope and objectives	Actors
Plug and play of smart appliances in house	This sub-use case aims to specify the smart appliances (load with some intelligence control capability and flexible demand) that have the plug-and-play functionality. These devices are not directly controlled by the microgrid control centre	Consumer and network smart devices
Registration of devices owned by the microgrid operator at the microgrid control centre	This sub-use case aims to ensure that any new or replaced device added by the microgrid operator is compatible with the microgrid standards. Therefore, the microgrid control centre should register, and specify the services and capabilities of such devices and reconfigure accordingly. In addition, the microgrid control centre should ensure the secure and trust channel for monitoring and controlling these devices	Microgrid operator, microgrid control centre, and network smart device
Registration of DER and storage units at the microgrid control centre	This sub-use case aims to provide the privacy and security of the DER and storage owners to be able to offer the services from their units. The microgrid control centre should build a trust channel to monitor and control the registered DER and storage units to provide the services to the microgrid and incentives to the owners. Access rights should be predefined based on contractual agreements	Microgrid operator, microgrid control centre, DER owner, and storage owner
Registration of a home/building energy management system at the microgrid control centre	This sub-use case aims to provide privacy and security of the communication link between any home/building management system in the microgrid at the microgrid control centre. The home/building management system aims to link all the smart appliances in the home/building with the microgrid control centre to receive the control and pricing signals accordingly. The home/building management system aggregates the capabilities of the smart appliances to reduce the complexity of monitoring and controlling these appliances. Access right should be predefined based on contractual agreements	Microgrid operator, microgrid control centre, home/building energy management system and network smart device

with the energy markets to manage and contract the offers and bids in the financial interactions of energy or ancillary services exchange. Whereas the "Control and Management" use case are describing the actions that can be taken to run a function of the devices, systems and/or sub-systems. Therefore, identifying the proper use case(s) and describing their main actors aim to reduce the complexity of proposing a set of trial scenarios, which can be evaluated based on the technical and regulatory requirements.

References

1. Eger K, Götz J, Sauerwein R et al (2011) Microgrid scenario building blocks. Munich, Germany
2. Werven MJN van, Scheepers MJJ (2005) The changing role of energy suppliers and distribution system operators in the deployment of distributed generation in liberalised electricity markets
3. Vesco A, Ferrero F (2015) Handbook of research on social, economic, and environmental sustainability in the development of smart cities
4. Iberdrola Distribucion Electrica (2009) Flexible electricity networks to integrate the expected "energy evolution"
5. Karg L, Kleine-Hegermann K (2011) Sustainable power systems for the alpine space: guidelines for regional decision makers and practitioners. Berlin
6. Losi A, Mancarella P, Mander S et al (2013) Recommendations for standard committees, regulators, stakeholders groups, future R&D
7. Degner T, Schmidt J, Strauss P (2005) DISPOWER—distributed generation with high penetration of renewable energy sources. Kassel
8. Soriano R, Alberto M, Collazo J et al (2011) Open node—open architecture for secondary nodes of the electricity SmartGrid. In: 21st International conference on electricity distribution. CIRED, Frankfurt, p 4
9. Australian Energy Market Operator (AEMO) (2019) Electricity demand forecasting methodology information paper important notice purpose

Chapter 3
Market Participations and Business Models in Microgrids

3.1 General

Microgrids have several actors that can contribute to the investment decision in designing, operating, and expanding such projects. The general classification of the actors can be determined based on the people and organisations that own the assets or are interested to invest in the microgrid area as well as the third-party bodies that operate the microgrids or represent some of the actors to manage the interactions and contracts. In addition, devices, systems, and sub-systems can be considered as main actors in microgrids, which are responsible for controlling, managing, and providing the support decisions in several actions. These main actors are also considered as main players in the energy markets as well as the business models. In this context, several energy markets are currently available (e.g., wholesale energy market), while some other markets are either operated in parallel with the existed markets or are expected to be coming in the near future. As the energy markets are changing based on the new technologies and tariff schemes; therefore, being familiar with these markets helps in selecting the proper use cases that can significantly guide the design and operation process to secure reliably and stably microgrids. Also, that can help in providing feasibly scenarios that ensures the higher benefits to the prosumers and operators. Thus, these aspects are significantly important in building a proper business model to engage the prosumers and attract more green investments in this field.

3.2 Main Actors in Microgrids

Several actors play a significant role in designing and operating microgrids as well as participating in the energy market. Those actors are the keystone of the business models. In addition, they appear in multiple use cases. Therefore, they should be clearly defined to be appropriately linked with the business models and use cases.

I. A. Ibrahim et al., *Microgrids Design and Operation*, Power Systems,
https://doi.org/10.1007/978-981-96-4914-3_3

Accordingly, the main actors in microgrids are categorised based on their roles into "people and organisations" and "systems and devices" [1–3].

3.2.1 People and Organisations

People and organisations actors represent the owners and operators of the microgrids who can be considered the players in the wholesale market, including the balancing and ancillary services. Those actors must be considered in designing and operating the microgrids to either run and or benefit from the services provided [1–3]. Accordingly, all the actors in the "People and organisation" category are defined in Table 3.1.

3.2.2 Systems and Devices

Systems and devices are also playing a significant role in managing the energy flow, stability, and reliability in microgrids. These systems and devices can be operated and controlled by microgrid operators, prosumers, and/or overlay grid operators [1–3]. Therefore, defining the systems and devices operator has a significant impact on extending the understanding of the design and operation of the microgrids. Here, all the actors in the group "Systems and devices" are described in Table 3.2.

3.3 Types of Energy Markets

In order to understand the different types of energy markets related to microgrids, it is important to know the following [1, 4–6]:

- Future (regional) retail energy market represents buying and selling energy from and to the external market. The local trading inside the microgrid is not covered here.
- General wholesale energy market includes the main local and external energy market, and the spot energy market. This market represents an open market for all the demand and supply sides to trade the energy for day-ahead and long-term as in the main energy market or short-term basis as in the spot energy market.
- Balancing and regulation market aim to ensure the balancing between demand and supply sides during any power shortages.
- Ancillary services market represents a pool of several ancillary services (e.g., frequency control, voltage support, etc.) that are required to improve the reliability of the distribution and transmission systems.

In this context, vertical integration (i.e., integration within one product column in contrast with horizontal integration) is a common structure of the energy market,

Table 3.1 Microgrid actors—people and organisations

Name		Type	Description
Microgrid operator		Organisation	A microgrid operator that operates the microgrid and manages the local pricing policy and trading with the electrical energy market. In traditional business models and small microgrids, microgrid operators can also be in charge of the roles of the energy retailers and aggregators
Microgrid operator	Retailer	Organisation	An electrical energy market player that sells the electrical energy to consumers on behalf of the distribution network operators (DSO) or transmission system operators (TSO)
	Aggregator	Organisation	An electrical energy market player that buys or sells electrical energy on behalf of two or more consumers, or DERs and storage systems owners
	System operator	Organisation	System operator that responsible to operate, providing maintenance, and if necessary, developing for the microgrid's distribution system
Prosumer		Person/ organisation	The term prosumer refers to consumers, DER, and storage owners
Prosumer	Consumer	Person/ organisation	A consumer that acts as a private, industrial, governmental customer or transportation system. Also, the consumer may operate controllable fixable loads
	DER owner	Person/ organisation	A DER owner that operates a DER connected to the microgrid
	Storage owner	Person/ organisation	A storage owner that provides storing and delivering the previously stored electrical energy services to the microgrid
Overlay grid operator		Organisation	A grid operator that operates the main grid before the PCC of the microgrid. The overlay grid operator is responsible for planning, constructing, operating and providing maintenance of the distribution and transmission electric power networks
Service provider		Person/ organisation	The service provider that assists the microgrid operator in running the microgrid by providing a variety of services (e.g., energy market analysis, load/generation profile forecasting, etc.)

Table 3.2 Microgrid actors—system and devices

Name	Type	Description
Microgrid control centre	System	It controls and manages the microgrid operator's activities, and the aggregation of supply and demand
Overlay grid control centre	System	It controls and manages the overlay grid where all necessary supervision and control duties are carried out
DER unit	Device	DERs include distributed generation (e.g., small PV and wind turbines) that is connected to the microgrid, and it can be monitored and controlled
Storage unit	Device	A power reserve provided by the storage unit that can be monitored and controlled
Home or building energy management system	System	It connects the smart home/building (e.g., controllable loads) with the microgrid control centre to provide the services of improve the operation of the microgrid
Network smart device	Device	The network smart devices are the intelligent devices that can be supervised and controlled at the feeder level (e.g., sensors, circuit-breakers, and switches)

and it refers to a company's ownership of its upstream suppliers and downstream customers. The main actors in the energy markets based on the vertical integration are given as follows [4]:

- **Energy supplier—large power producer**. A large power producer is an energy provider capable of meeting the short-term and volatile demand of domestic customers in a cost-effective manner. Because of its own production facilities, the large power producer is less reliant on the balancing and wholesale market.
- **Energy supplier—DER operator**. The advantages of a DER operator are nearly identical to those of a large energy producer, with the exception that DER generates power on a smaller scale than a large energy producer, making it less applicable. Because of the environmental benefit, the government's policies and subsidies are more favourable for DER operators. DER operator could be the DER owner or the microgrid operator.
- **Consumer**. Consumer represents private, commercial/industrial, or transportation system energy demand and can purchase the energy for their own use based on their preferences according to the energy market prices and offers (e.g., tariff structure).
- **Consumer—DER operator (partial producer)**. The on-site electricity generation is consumed by the consumer, who is the owner and operator of the DER. In this integration, the consumer bypasses the energy supplier's sales process, saving the network cost.
- **DSO**. DSO is responsible to operate the distribution systems to meet the demand of the distribution loads with long-term reliability, ensure that the system services (e.g., power quality) are meeting the requirements, and offering the needed maintenance and infrastructure upgrade.

- **TSO**. TSO is responsible for operating and providing the proper maintenance of the transmission systems and coordinating the interconnections with other systems to meet reasonable transmission of electricity demands. Several other services should be offered by TSO to ensure long-term reliability, such as reserve capacity, power quality, balancing services, reactive power supply, etc.
- **DSO—TSO coordination**. In case the DER penetration is high, the DER operator necessitates a great deal of coordination between TSO and DSO. The integration of both solves the coordination issue.
- **DSO—DER operator**. The DSO can benefit from DER's local ancillary services. The placement of DER on optimal sites is critical for coordinating with DSO and meeting their demands.
- **DSO—energy supplier**. The simple flow of transactions from customer to energy supplier and then to the DSO is an advantage of DSO integration with energy suppliers. The disadvantage is the difficult exchange of information between both, such as metering data.

3.3.1 Future (Regional) Retail Energy Market

The retail energy market is a new concept of energy markets which could be formed in the near future. This market can be defined as a centralised market operated by a third party (e.g., e-Boker offers online brokerage electrical services). This market can be connected to the wholesale market to allow an open energy trading between microgrid operator and/or prosumer with the external market (internal trading into the microgrid is not considered) through the overlay grid operator. Alternatively, microgrid operator can represent all the prosumers in the external market and manages the internal demand/supply balancing inside the microgrids in the best way to ensure the reliability and stability of the microgrid as well as securing the high financial benefits for the customers. The structure of this market allows buying and selling the energy and ancillary services [7].

3.3.2 General Wholesale Energy Market

This market is representing the current energy market, where it is opened for all the demand and supply sides to buy and sell the energy based on pre-defined contracts and/or tariff structures [4]. Some of the large power producers can have a dynamic tariff structure to sell the energy inside the microgrid and/or to the external market. In Australia, the retailers are responsible in putting these contracts and tariff structures. Some energy schemes can also be followed to manage buying the energy from the DER owners, such as feed-in tariff (FiT) scheme or others. The overview transactions in the wholesale energy market (today's electricity market) can be illustrated as in Fig. 3.1 (numbers representation is explained in Table 3.3).

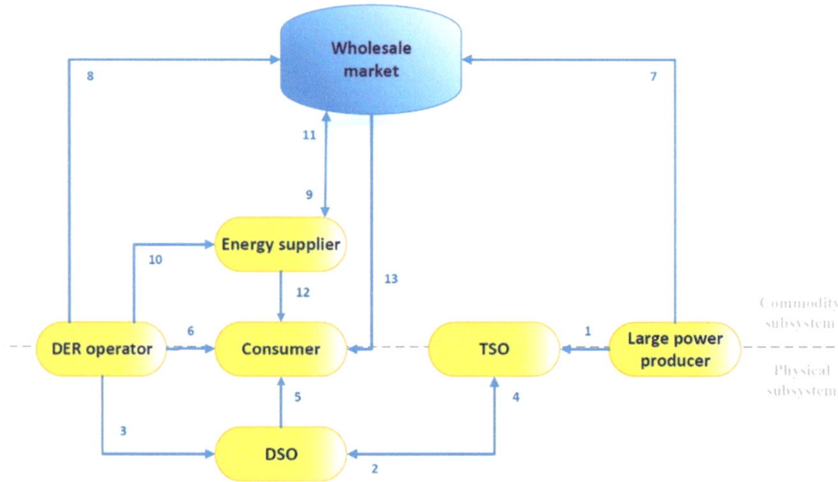

Fig. 3.1 Overview of the electricity and economic transactions within wholesale energy market

The energy market has main two subsystems, which are representing the physical hardware to generate and transport the electricity through the main grid to the customers (i.e., physical subsystem), and the platform to support the prosumer (e.g., DER owner, customer) activities (e.g., DER owner generation, energy trading, and energy consumption) and their mutual relations (i.e., commodity subsystem). Based on Fig. 3.1 the financial transactions and information exchange between the main actors within the wholesale energy market to allocate the costs and benefits can be summarised based on the numbers order in Table 3.3.

3.3.2.1 Main Energy Market

This market is a sub-market from the general wholesale energy market, which is representing the financial transactions (i.e., offers and bids) nationally to provide the most suitable electrical energy in the national energy market considering the business aspects. This market offers open trading between all the demand and supply sides in day-ahead and long-term plans. This type of market is mostly operated by the DSO and/or retailer.

3.3.2.2 Spot Energy Market

The spot energy market is also a sub-market from the general wholesale energy market, which is representing an open trading market to provide offers and bids in short-term plans (i.e., seconds to a few minutes (e.g., 5-min in Australia)) to

Table 3.3 Transactions within the wholesale energy market

No.[a]	Actor		Offers	Expects in return
	From	To		
1	TSO	Large power producer	Connection to and use of transmission network	Connection charge, and in some jurisdictions use of system charge
2	TSO	DSO	Transmission network-distribution network power transmission services	Use of system charges on behalf of their customers
3	DSO	DER operator	Connection to and use of distribution network	Connection charge, and in some jurisdictions use of system charge
4	TSO	DSO	Distribution network—transmission network power transmission services	Use of system charges on behalf of their customers
5	DSO	Consumer	Power distribution services	Transmission system and distribution system charges
6	DER operator	Consumer	Power (physical subsystem)	Full cost coverage (auto production)
7	Large power producers	Wholesale market	Power (commodity subsystem)	Contract price
8	DER operator (consolidator)	Wholesale market	Power (commodity)	Contract price
9	Wholesale market	Energy supplier	Power (commodity subsystem)	Contract price
10	DER operator	Energy supplier	Power (commodity subsystem)	Contract price (often based on a power purchase agreement)
11	Energy supplier	Wholesale market	Power (commodity subsystem)	Contract price
12	Energy supplier	Consumer	Power (commodity subsystem)	Contract price
13	Wholesale market	Large consumer	Power (commodity subsystem)	Contract price

[a] No. reflects the numbers shown in Fig. 3.1

customers and large power producers. This market can be run through third-party organisations or governmental agencies (e.g., AEMO) [8].

3.3.3 Balancing and Regulation Market

The balancing and regulation market is a fact of liberalisation of energy markets in many European counties, where it is available in parallel with the general wholesale energy market. This market aims to provide balancing services (when the supply meets the demand) to stabilise the electrical networks [4]. In this market, the TSO is operating and managing the energy flow to achieve the energy balancing in the distribution networks and microgrids. TSO is considered as the only buyer; however, the large energy producers and microgrid operators (as representative of DER operators) have access to such a market. In this market, most of the offers are carried out by the large energy producers, while TSO can charge the energy suppliers for causing the imbalance. In contrast, TSO accepts high bids in case of any electrical power surplus. These financial transactions are arranged in predefined contacts between all the players in this market. Here, the balancing market prices (i.e., imbalanced prices) are mostly higher than the normal energy prices as higher marginal costs are considered to avoid any large electrical power generation does not comply with the pre-defined contracts. This leads to encourage market players to forecast the electricity production and demand as accurate as possible and to act in accordance with these energy programs [1]. In this context, Table 3.4 shows the balancing and regulation services that can be provided to the grid. In addition, the electrical energy transactions into the balancing and regulation market within the electricity market is illustrated in Fig. 3.2.

Table 3.4 Balancing and regulation services

Service	Definition	Nature	Providers
Primary regulation	All the generators regulate their output automatically to mitigate the frequency changes by less than 30 s	Unpaid and mandatory	All generators
Secondary regulation	Some generators regulate their outputs to avoid changes in frequency within 100 s	Paid and facultative	Some generators that authorised by the network operator within the regulation areas
Tertiary regulation	Some generators and pumping units regulate their performance to response to power variation within 15 min and remain producing power for at least 2 h duration	Paid and facultative (mandatory offer)	Some generators and pumping units enabled by the network operator within the regulation areas

Fig. 3.2 Overview of electricity and economic transactions within the wholesale energy market, including the balancing and regulation market

3.3.4 Ancillary Services Market

The ancillary services market is not a standalone market currently (it may be coming up as a separate market in the near future) and it is operated in parallel with the balancing and regulation market within the wholesale electricity market. This market can offer several services by the power producers and DER operators to the main grid (i.e., overlay grid operator) to improve its reliability and ensure the power quality. These services can include compensation of the power losses, mitigate the frequency issues through local and/or coordinated frequency control strategies, mitigate the voltage issues through controlling the import/export limits of the active and/or reactive powers or using voltage regulators, and restoration of supply services, such as temporary island operation, and black start [4]. The ancillary services provided by the DER operates could be relatively low, but it could be beneficial once the constraints (e.g., terrain-related constraints, planning and environmental (e.g., carbon dioxide (CO_2) production control)) limit the network development. The possible ancillary services that large power producers and DER operators can provide at the distribution level are listed in Table 3.5. In addition, the electrical energy transactions by including the balancing and ancillary services markets within the wholesale energy market can be as shown in Fig. 3.3.

Summary

Microgrids have several benefits that attract the electricity operators, governments, and private investors to sponsor several projects around the country and

Table 3.5 Possible ancillary services provided by large power producers and DER operators

Ancillary service	Large power producer	DER operator
Compensation for power losses	Capable to do it	Capable to do it
Frequency control	Capable to do it	Able to do it based on the type and capacity of DER
Voltage support (volt/watt control)	Able to do it, but it needs more investments from DSO for reinforcements and extensions	Capable to do it
Reactive power control	Able to do it, but it needs more investments from DSO for reinforcements and extensions	Capable to do it
Black start	Capable to do it	Capable to do it
Reserve	Capable to do it	Capable to do it

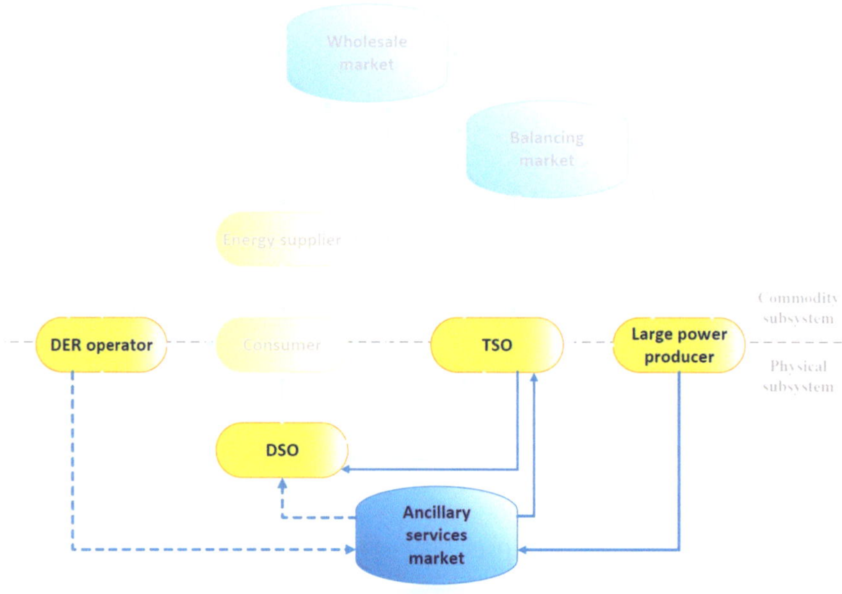

Fig. 3.3 Overview of electricity and economic transactions within the wholesale energy market, including the balancing and ancillary services markets

worldwide. These benefits contain technical improvements, economic returns and environmental impacts. The success of designing, implementing and operating microgrid projects strongly depends on the social acceptance, which helps in engaging the prosumers to invest in such projects to expand the services and reduce the costs. Several actors are involved in this process. Thus, defining all the main actors and understanding their responsibilities and needs can significantly contribute to linking them with the energy markets as key players. This leads to extending the understanding of the services that they can provide to the energy markets as well as the benefits they will gain. Accordingly, the understanding of the operation of general energy markets (e.g., wholesale energy market) and mapping it with the services markets can extend the prosumer knowledge and trust related to the financial benefits and the offered incentives, which can increase their engagement and significantly increase their investments to extend their contribution towards more microgrids.

3.4 Microgrid Business Model

Business model for microgrids is defined as "the way in which a microgrid project or business is planned, implemented, and executed to meet strategic objectives" [9]. This section discuses two types of business models. The first type shows the influence of DERs in microgrids on the financial interactions of the DSO and energy suppliers. The second type is more specified for the managing the financial iterations, roles, right and ownerships within the microgrids.

3.4.1 Business Models of the DSO and Energy Suppliers

To assess the influence of DER on the business models of DSOs and energy suppliers, the financial interactions, including the revenues and expenditures explained through the business models of energy suppliers and DSOs in the current market [4].

3.4.1.1 Business Model of DSO

The DSO business model aims to manage the revenues, and expenses within the microgrid considering the network charges for DER operators. Accordingly, the block diagram representation of the business model of DSO is represented in Fig. 3.4.

Accordingly, the primary sources of DSO's revenues and expenses are listed in Table 3.6.

Network charges for DER operators can be based on two types of cost billing, which are detailed below:

- **Shallow cost charging**. This type of billing implies that the DER operator can pay for the connection line closest to PCC.

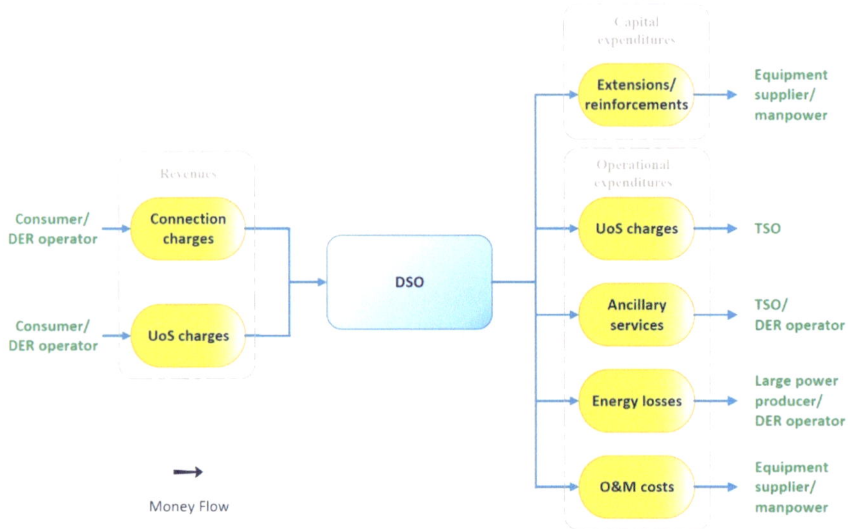

Fig. 3.4 Business model of the DSO

Table 3.6 DSO's revenues and expenses

Revenues	Expanses
• The used electricity charges from consumers and the DER (in kWh and/or kW) • Both consumers and DER operators pay network cost	• Capital expenditures on network and distribution infrastructure such as transformers and grid extensions, among other things • The DSO bears all operational costs, including network maintenance, consumed power costs to TSO, ancillary services to TSOs and DER operators, and distribution system losses

- **Deep cost charging**. This form of billing implies that the DER operator can pay for both the connection line and the infrastructure required for network upgrades.

In general, the tariff structure of the network is based on the cascade concept, which means that the consumer not only pays for the network level to which the belong premise is connected, but also for all higher network layers in relation to the time they spend on these network layers. Based on the amount of electricity flowing towards the maximum load, the DSO compensates the TSO.

Several factors can affect the current business model of DSO. These factors are listed as follows:

- **Increase in DER penetration**. Increased customer installation and use of DER poses a severe threat to DSO's economic model because DER is located closer to consumer demand and only requires minimal transportation expenses for electricity supply. When the consumer entirely utilises, the on-site electricity provided by the DER's autonomous operation, the DSO's revenue will not be existed if the operator does not use the DSO infrastructure. Secondly, increasing DER penetration raises technical issues that necessitate modifications in distribution system design and management for stable and balanced distribution network operation.

- **Regularity incentives**. The regulation of the electrical grid is critical in ensuring that consumers have access to the grid. Electricity transportation is a common monopoly, and if the utility grid's owner has an interest in other aspects of electricity, such as energy supply or production, they can blackmail the DSO, energy provider, or power producer by raising the grid's entry barrier. To lessen the threat to DSO's business model, the interests of utility grid must be totally unbundled through incentive regulation to avoid distortion, competition, and favouritism.
- **Connection charges for DER**. If DER penetration is high in a specific place, network connection prices may soar, necessitating network upgrades to accommodate the flow of electricity efficiently. In this situation, DSO views DER as a threat since DSO finds it impossible to pass on undisclosed reinforcement costs to customers due to efficient incentive regulation.
- **Unbundling**. DSOs are generally not permitted to be power producers since there is a strict unbundling boundary between DSO, power producer, energy supplier, and grid. The benefit of DSO owning producing capacity is that DSO's line losses, operating and capital expenditures can be decreased. The unbundling problem can be overcome by paying the DER operator a monetary incentive to control the DER unit during periods of high consumer demand. Because the DER's lifespan is short and the DER operator can discontinue operations at any time owing to economic problems, the financial contract with the DER operator poses a threat to DSO.
- **Behavioural strategies of DSOs**. There are three types of behavioural strategies that can be adopted by the DSOs during the adaptation process in case of competition and accompanying regulation. These behavioural strategies are given as follows:

  - **Stabilisation strategy**. In this strategy, the main goal of the DSO is to reduce the uncertainty in the whole electricity sector.
  - **Defensive strategy**. The DSO becomes aware of any negative consequences of rapid DER growth and restrictions on its business as a result of this strategy. The DSO is putting up a straggle against the regulatory process that is reducing their profits.
  - **Entrepreneurial strategy**. Instead of resisting regulation, the DSOs use this strategy to explore for other alternatives by developing new regulatory structures and initiatives to mitigate the DSO's business model threats.

3.4.1.2 Business Model of Energy Suppliers

In the wholesale market, the energy supplier purchases electricity from the power producers and then sells it to the consumers. Here, the business model representation of the energy supplier is illustrated in Fig. 3.5.

Accordingly, the primary sources of energy suppliers' revenues and expenses are listed in Table 3.7.

The other factors that influence the business model of electricity supplier are given as follows:

- The energy supplier may also own manufacturing capacity in order to optimise the energy demand and supply cycle.
- Purchasing power from infrequent sources with fluctuating power may increase the balancing cost.
- The energy supplier may be eligible for government incentives for purchasing and selling green energy.

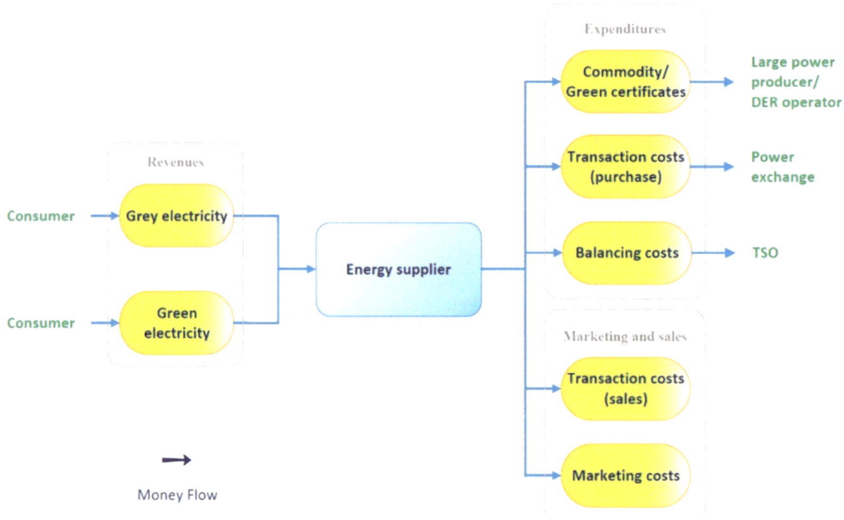

Fig. 3.5 Business model of the energy supplier

Table 3.7 Energy suppliers' revenues and expenses

Revenues	Expanses
• Sales of electricity to consumers at retail prices • Providing consumers with various energy/non-energy related products, services, and technologies	• Purchasing electricity from generators • The TSO's balancing charges, if any • The cost of marketing • The cost of purchasing and selling electricity, including the cost of access to the power exchange

3.4.2 *Business Models for Utilities*

Microgrids have long been viewed as a systemic threat to utilities' income streams. As more consumers become self-sufficient, utilities are compelled to raise the cost of power delivered to compensate for lost revenue. As a result, customers are even more enticed to leave the grid, resulting in the so-called "utility death spiral." As a result of this problem, utilities are adjusting their traditional business models to suit the growing use of microgrids in MV and LV distribution networks [10].

Accordingly, utility business models for microgrids are divided into two types. First, regulated utilities may become "smart energy integrators" [11]. Utility companies would no longer be able to own generation assets and would no longer be in the business of selling power. Instead, they would act as supervisors for energy and information, maintaining the optimal balance between supply and demand, while engaging customers and implementing various smart grid programs. In this case, customers freely select private or public power suppliers. Second, utility companies might become "energy service providers" by exploiting new technologies and expanding their current business from selling power to selling control services in addition to other forms of energy. In this scenario, utilities may continue to own generation assets and sell directly to customers [10]. In this context, four utility microgrid business models have been developed and deployed to yet [9], which can be summarised as follows:

- **Rate-base**. Rate-base is defined as the value of property on which a utility is entitled to earn a particular rate of return as set by a regulatory authority. As a result, the utility's income is set by the rate-base and the allowable rate of return, while the expenditures are passed on to customers. Thus, under this business model, the utility implements microgrids and, in exchange for increased electrical supply resiliency and reliability, incorporates the expenses of developing and managing them in its base-rate. While this model is not prevalent yet, promising steps have been taken by some North American utilities.
- **Operation and maintenance (O&M)**. As the penetration of microgrids increases, the demand for O&M services will generate a new utility business model. A business model like this allows utilities to generate a steady income stream, either fixed or variable, from microgrid deployments in their service regions. Utilities in their jurisdiction can provide O&M services to microgrid owners.
- **Design, build, own, operate, and maintain (DBOOM)**. DBOOM, as the name implies, is a business model in which the utility is in charge of all aspects of microgrid design, implementation, operation, and maintenance. Utilities may use the DBOOM business model instead of infrastructure changes to provide consumers with greater supply resilience and reliability.
- **Public–private partnerships (mixed ownership)**. The DBOOM business model presents a difficulty in that a utility must provide a wide variety of hardware and software services while remaining agnostic to a varied group of microgrid suppliers and legacy systems. Partnerships with third-party suppliers that provide such services, on the other hand, can be a critical enabler. This type of strategy is

Table 3.8 Advantages and disadvantages of utility microgrid business models

Business model	Pros	Cons
Rate-base	Ideal for communities with low reliability of supply	High capital expenditure, higher supplied cost of energy for customers
O&M	No capital expenditure needed	No influence on microgrid control, low revenue
DBOOM	Full control on the assets and revenue stream	High capital expenditure, lack of expertise in-house
Mixed ownership	Low capital expenditure, moderate need for in-house expertise	Revenue may be limited, limited need for in-house expertise

known as "public–private partnerships". A public–private partnership, according to the definition, "is a contractual arrangement made between a public agency and a private sector company that allows for increased private sector engagement" [12]. In this arrangement, utilities refer their customers to a commercial vendor who may offer microgrid projects' hardware and/or software component.

The several utility microgrid business models that described above have benefits and drawbacks, which can affect the selection and implementation based on the stakeholder decisions, use case(s) and the main actors in the microgrid. Here, the benefits and drawbacks of the microgrid business models are summarised in Table 3.8.

Summary

DNSPs are developing innovative business models to position themselves as a prominent partner in various types of microgrid ownership and economic frameworks. To build a comprehensive business model, local resources and accessible technology must be assessed, and expected services and operational objectives must be established in advance. Consequently, given the techno-economic environment of DNSPs operation, an optimal technology set may be chosen. Another business model component is the ownership, revenue, and financial structure. A microgrid can be owned by a single entity, such as the utility or prosumer, or have a mixed ownership. Additionally, the revenue generation model is dictated by several factors, such as DNSP regulations, the microgrid integration eco-system, retail market competition, and the service territories. Therefore, selecting the proper use case(s) based on the stakeholder decisions, and the main actors' engagement play a significant role in representing the business model that can lead to achieve the near and long-term goals. The most used business model based on the national and international projects is DBOOM business model.

References

1. Eger K, Götz J, Sauerwein R et al (2011) Microgrid scenario building blocks. Munich
2. EU Commission Task Force for Smart Grids (2011) Expert group 3: roles and responsibilities of actors involved in the smart grids deployment
3. Mander S, Ghanem DA (2013) Key societal factors influencing the adoption of the ADDRESS smart grids architecture
4. Werven MJN van, Scheepers MJJ (2005) The changing role of energy suppliers and distribution system operators in the deployment of distributed generation in liberalised electricity markets
5. Vesco A, Ferrero F (2015) Handbook of research on social, economic, and environmental sustainability in the development of smart cities
6. Alagna V, Cauret L, Entem M et al (2011) Description of market mechanisms which enable active demand participation in the power system
7. ICCS, PPC, GERMANOS et al (2005) More microgrids: investigation of regulatory, commercial, economic and environmental issues. http://www.microgrids.eu/micro2000/index.php?page=tasks&id=7. Accessed 15 Mar 2022
8. AEMO (2020) Integrated system plan for the national electricity market
9. EPRI (2020) EPRI's energy storage roadmap, vision for 2025. Palo Alto, California
10. Sub-Committee on MicroGrids Whitepaper, IEEE Smart Grid Technical Activities Committee (2022) Microgrids: utility challenges and opportunities
11. Andreini D, Bettinelli C, Foss NJ, Mismetti M (2021) Business model innovation: a review of the process-based literature. J Manag Gov. https://doi.org/10.1007/s10997-021-09590-w
12. Farrokhabadi M, Canizares CA, Simpson-Porco JW et al (2020) Microgrid stability definitions, analysis, and examples. IEEE Trans Power Syst 35:13–29. https://doi.org/10.1109/TPWRS.2019.2925703

Chapter 4
Modelling and Economic Analysis of Microgrids

4.1 General

Modelling of the microgrids has a variety of purposes. The main purpose is to understand the behaviour of the several components and mimicking their actual performance (or approximate performance with an acceptable error range) to provide the optimal design and operation scenarios of the microgrids. In addition, it allows the "what-if" scenarios. Moreover, the modelling of the microgrids can also aims: (i) to find out what might happen in the present and the future (e.g., what would happen if a new producer or consumer is added. The effects on other consumers and producers can also be examined); (ii) to analyse the energy flow in the network (i.e., energy flows between nodes of the network, resulting in the solution of the simulation problem); (iii) to analyse the effects of defects and breakages in a line; and (iv) to adapting various models so that modelling the network differently to solve the problem before it happens [1].

The modelling cycle, which can be used to model the microgrids is illustrated in Fig. 4.1. As a starting point, a simple model can be created and simulated. The obtained results of the simple models can be validated with actual data. After assessing the model accuracy, this model can be executed and utilised if it is suitable. When the model is not suitable, it is improved, and a more complex one can be created to consider more parameters. Then, the simulation steps are repeated and verified again whether it fits the actual performance. If it fits, the model will be used. Otherwise, it can be continued with the next modelling cycle until it is satisfied.

The modelling cycle depends on the application and the availability of the data. A simple model may be sufficient for specific applications, whereas, for other applications, a very complex model can be required to describe reality adequately. For example, it is not necessary to provide a detailed description for expansion planning, as only the general trends are required (e.g., increasing consumption rates). Consequently, it is sufficient to have only a global overview and not a detailed model. Thus, grid planning generally calls for a simpler model. On the other hand, for the system

I. A. Ibrahim et al., *Microgrids Design and Operation*, Power Systems,
https://doi.org/10.1007/978-981-96-4914-3_4

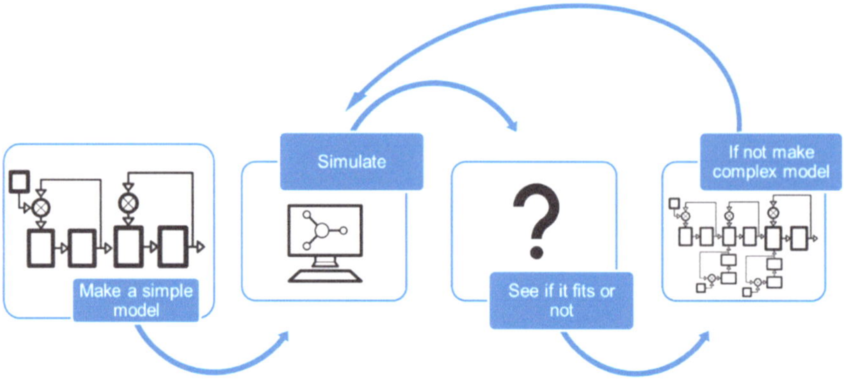

Fig. 4.1 General modelling cycle

operation, the impact of an event on the grid dynamics should be investigated. In this circumstance, it is necessary to know what will happen, not only at the current status but also at the future status (i.e., in 5 or 10 min, or so). Therefore, more complex models are required to express the behaviour of the system accurately.

4.2 Microgrid Components

Microgrid components can comprise feeder/network (i.e., multi-feeders), loads (e.g., controllable, non-controllable and critical), DERs (e.g., dispatchable DERs and non-dispatchable DERs), ESS, EVSEs, and other components (e.g., electricity and communications networks devices, protection devices, etc.). Beside these components, microgrids need more infrastructure, such as PCC (could be represented by circuit breaker, solid-state switch, or back-to-back converter), controllers (for energy management, switching and stability actions), and monitoring devices (for sensing and measuring the network and components status and data (e.g., voltage, frequency, etc.) and send it to take the proper control action and/or update the characteristics of the system). The general structure of microgrid components that are managed by the microgrid control system is illustrated in Fig. 4.2.

4.2.1 Network Topology

Microgrid could be integrated in one feeder or multiple feeders (i.e., network). The factors of designing the microgrid in on feeder or in multiple feeders are the use case(s), load sensitivity, level of reliability, and project budget. Accordingly, the microgrid network topology could be radial, ring or mesh. Each of these topologies

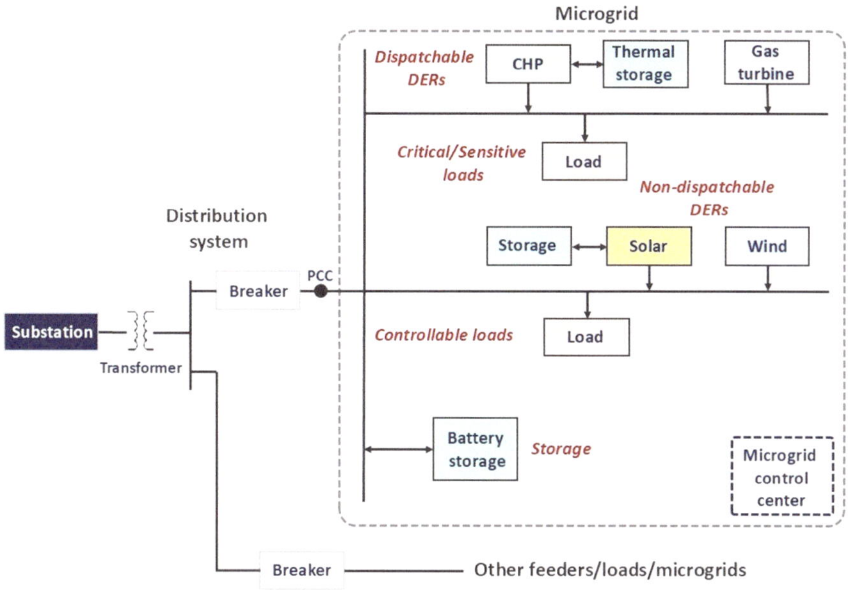

Fig. 4.2 General structure and components of a microgrid

has advantages and disadvantages as well as cost considerations. Radial topology is the most popular microgrid architecture, due to the simply of its implementation and suitability for most applications and use case(s). Also, it is the cheapest option compared with other architectures. However, it has relatively low reliability than the other architectures as it does not have any backup feeder to operate in case of any internal fault happened. While ring topology has two feeders to connect the DERs with the loads; therefore, it could operate on one feeder if the other feeder faced an emergency or fault. Although the reliability increased compared with the radial topology, the cost is also increased relatively. Some of the microgrid applications are to secure a high reliability and 24/7 energy supply to the loads, especially if all the loads are sensitive such as in military microgrids. Thus, mesh topology can be used in such applications as its design has more than two feeders, which increases the reliability significantly as well as the implementation and operation costs [2]. In Australia, most of the electricity distribution networks and microgrids are operated in a radial topology (no "loops") [3].

4.2.2 Electrical Loads

Loads can refer to all of the microgrid's components that convert electricity into various types of energy for end usage. Electrical loads can be categorised based on their natural as resistive (e.g., lamps and heaters), inductive (e.g., transformers, generators, motor, etc.), capacitive (e.g., three-phase induction motor starting circuit) or some combination between them. The electrical loads have several types based on the application and size, such as residential, commercial, industrial and agriculture loads. In this context, loads can be categorised based on their sensitivity nature into sensitive/critical and non-sensitive/non-critical loads. The sensitive/critical loads cannot be controlled (e.g., decreased, shifting or shaving) and/or curtailed under any condition. While the non-sensitive/non-critical loads can be controlled based on the energy management and stability actions under demand response program in grid-connected mode and/or disconnected in the islanded mode to guarantee the supply to the critical loads [2]. The loads can also be categorised based on the possibility of controlling them into controllable and non-controllable loads. Controllable loads are considered as non-sensitive/non-critical loads as well as they are flexible and have programmable functions, where they are able to be controlled directly or indirectly (e.g., through dynamic electricity pricing) to adjust their base load profile in response to a set of operating constraints. Some examples of flexible and programmable loads could be electrical heaters, air conditioning systems, refrigerators, and water heaters. It is worth mentioning that active control strategies (e.g., volt/watt and volt/var) can be implemented for these loads [4]. In contrast, the non-controllable loads are either sensitive/critical loads or they are not flexible and not supporting the programmable functions.

4.2.3 Distributed Energy Resources

DERs are the main energy generators in microgrids, which can be classified as dispatchable DERs and non-dispatchable DERs. The dispatchable DERs are the generators that their output can be fully controlled. The dispatchable DERs contain renewable and non-renewable energy resources. The renewable dispatchable DERs are including small-scale hydropower plants (e.g., small, mini and micro-hydropower). While the non-renewable dispatchable DERs are including small reciprocating internal combustion engines (e.g., diesel, gas, and biofuel from biomass and biogas), and microturbines. On the other hand, the non-dispatchable DERs are the generators that mostly have a variable output based on the meteorological variables and weather conditions [2]. The non-dispatchable DERs are mostly intermittent renewable energy resources, including solar photovoltaic (PV) systems and wind power systems. DERs can also be sub-categorised based on their grid-connection interface as directly connected synchronous-generator-based, directly connected induction-generator-based, directly connected doubly-fed-induction-generator-based, DC type with DC-to-AC converter interface, AC-type with AC-DC-AC converter interface [4].

4.2.3.1 Dispatchable Distributed Energy Resources

Dispatchable DERs are fully controlled generators, which capable of providing a firm capacity within less than 30 s and can meet microgrid's 2-h capacity requirement as an operating reserve capacity. Accordingly, dispatchable DERs are essential in any microgrid project, especially if the capacity of the storage batteries is not enough to provide the capacity requirement either for their not existence in a microgrid design or due to the financial limitations. Thus, dispatchable DERs can minimise the investment costs as the size of the storage batters will significantly decrease. Dispatchable DERs can also offer the black start service of the microgrid in the islanded mode and ensure balancing the demand-side of the critical/sensitive loads. Here, the renewable and non-renewable dispatchable DERs are reviewed.

Small-scale hydropower plants

Small-scale hydropower plants can be categorised based on their capacity into small-hydropower (less than 10 MW), mini-hydropower (less than 1000 kW) or micro-hydropower (less than 10 kW) plants. Small-scale hydropower plants can be considered as a cost-effective solution in microgrid applications [2]. The principle of work of the small-scale hydropower plants is converting the energy of flowing water into electricity. The governance system of the small-scale hydropower plant does not differ much from the governance system of large-scale hydropower plants. The output power frequency is controlled by controlling the water flow through gates controlled by the regulator. The frequency of the output power must be adjusted to the microgrid frequency. The automatic voltage regulator controls the output voltage amplitude, which regulates the output voltage to match the microgrid voltage. When excess electricity is available in the microgrid, the generators used in the hydroelectric power stations can also be used as a motor to pump the water up to the hilltop, if existed [5].

Small reciprocating internal combustion engines

Small reciprocating internal combustion engines are well-known technology, which can operate using several types of fuels, including natural gas, diesel oil, petroleum, and biofuels. The reciprocating internal combustion engines can transfer the power to a shaft to generate the electrical energy using the piston-cylinder principle [2]. Several gas combustion technologies have been widely deployed due to the improvements in the electrical efficiency of the turbines, such as single cycle gas turbines (SCGTs) and combined cycle gas turbines (CCGTs). These new technologies are decreasing the levelised cost of energy (LCOE) [6].

Biopower technologies are generating the biofuels either from solid organic matter, which is animals or plants remnants (i.e., biomass) or from the decomposition of organic matter through microorganisms to digest these materials to generate gas. Direct combustion is used to generate the electricity by flowing the generated high-pressure steam, which is resulted from the biofuel burning process, over a series of turbine blades, causing them to rotate [7]. The output power from the biopower technologies strongly depends on the generated biofuel. Therefore, the availability of organic matter is a key source of such generators, which can be preferable to be used in microgrids that have access to farming areas. It is worthy to mention that these technologies are not 100% renewable as CO_2 is released in both technologies by burning material in biomass technology and when biogas is released from the decomposition of organic material in biogas technology.

Microturbines

Microturbines are small combustion gas turbines that can produce between tens to hundreds of kW (around 25–500 kW). It has a small size as the compressors, turbines, and generators are mounted on a common shaft. The microturbines can be operated using natural gas, biogas, or liquid fuels. These turbines are characterised to provide a high power generated to weight (or size) ratio as they have high rotation speeds (i.e., between 60,000 and 120,000 rpm [8]). In addition, they have several other features, such as lower emissions and noise compared with reciprocating internal combustion engines [2]. Usually, such turbines are used in cogeneration and trigeneration technologies, where it is needed to generate electricity, heat and/or cooling, to achieve higher energy efficiency [9].

4.2.3.2 Non-dispatchable Distributed Energy Resources

Non-dispatchable DERs are not flexible as dispatchable DERs in terms of turning them on or off to meet the demand at all time periods. In addition, they are considered as intermittent sources, which their output is fluctuating, and it is not always available due to its dependence on a factor that cannot be controlled (i.e., weather). Several types of non-dispatchable DERs are available such as small-scales tidal and wave powers. However, solar and wind powers are the most non-dispatchable DERs contributing to the distribution systems.

Solar photovoltaic (PV) systems

The output power from solar PV technology is intermittent and most often is considered as a non-dispatchable DER. The output power of the PV systems fluctuates and strongly depends on several meteorological variables, especially solar radiation, and ambient temperature, which are location depending. Solar PV technology converts sunlight directly into electricity. Its output is DC. This form of energy must be converted to AC before being connected to the AC bus in the microgrid. PV systems typically consist of a PV array and an inverter. The power-voltage curve of solar PV modules exhibits non-linear behaviour and may have multiple peak points depending on shading [5]. To ensure the operation of the PV system at the peak points due to the non-linear behaviour, software-defined algorithms called the maximum power point tracking (MPPT) algorithm are used. The goal of the MPPT algorithm is to operate the PV system with output DC power, voltage and current close to the point where maximum power is reached under a given environmental condition to match inverter inputs requirements [10]. The PV systems with rating capacity of a few kW to hundreds of kW can be installed on residential or commercial buildings. However, the solar PV farms can be rated at several MWs. PV systems are one of the most used DERs in microgrids, which are suitable for remote and islanded applications.

Wind power systems

Wind power is considered a highly intermittent and non-dispatchable DER due to the power output from the wind turbines is dependent on many factors (e.g., wind speed, air density, turbine characteristics, etc.). In addition, all of these factors are changing based on geographical location. Wind turbines convert wind energy into electrical energy. The wind turbines with a rated capacity of less than 75 kW can be considered as small-scale wind turbine based on IEC 61400-2 [11]. The small-scale wind turbines can be classified as horizontal and vertical axis wind turbines, which they can be installed in the rooftops or the gardens [2]. A wind turbine generator consists of a turbine blade and rotor, a gearbox, and an electrical generator. Wind energy is converted via turbine blades into mechanical power (torque) via a low-speed shaft, gearbox, and high-speed shaft that is then transmitted to a generator. According to the generator type, wind turbines may generate DC, AC, or variable frequency power output [8]. The induction generator is very inexpensive and easy to construct compared to the variable frequency generator. The disadvantage of the induction generator coupled to the wind turbine is that it forces the wind turbine blades to rotate at a constant speed, while the air speed is inherently variable. This disadvantage reduces the efficiency of the system. Thus, variable frequency synchronous generators can be used, which convert the variable air speed into variable frequency power. The power electronic converter (e.g., AC/DC/AC power converter) connected to the variable frequency generator converts the variable frequency to a static frequency that is tuned to the grid [5].

4.2.4 Energy Storage Systems

The energy storage systems include all the individual components or arrays used to store energy to defer the generation of electricity, including electromechanical (e.g., secondary batteries and flow batteries), chemical (e.g., fuel cells), electrical (e.g., capacitive energy systems), mechanical (e.g., flywheel, pumped hydro and compressed air), thermal (e.g., latent heat storage, sensible heat storage and thermos-chemical sorption storage), and hybrid storage systems [2]. The storage units and their electronic converters are key elements of the microgrid because they enable some ancillary services to be delivered, such as peak shaving, increase network asset utilisation, voltage support, and frequency control, notably when operating under islanding mode with variable renewable energy sources. Batteries are also necessary to successfully deploy local black start techniques [4]. Thus, storage systems can also be considered as a type of dispatchable DERs. The future states of energy storage, including the gaps that need to be filled to enable this future vision [12].

4.2.4.1 Battery Energy Storage System

Battery energy storage system (BESS) stores excess energy from a generation source, allowing the load to be supplied when the source is unavailable. Several technologies of BESS are available in the market, including lead-acid and advancing lead-acid, sodium sulfur, sodium nickel chloride, nickel–cadmium, lithium-ion, and flow batteries [2]. The charging and discharging process are mostly controlled by charge controller to protect the battery from overcharging/undercharging issues and to keep tracking the SOC. BESS plays an important role in controlling the microgrid frequency in the short-term in case of a disturbance in the microgrid. The response of most DERs integrated into the microgrid is slow. In the event of a disruption in the microgrid, BESS can act as a dynamic frequency stabiliser for a short period of time until the DERs responds to the disruption. The batteries used in the BESS cannot be charged quickly compared to the discharge rate. This means that BESS can accelerate rather than decelerate the energy system [5]. Thus, batteries are necessary to successfully deploy local black start techniques. To follow load variations in the microgrid and compensate for disturbances, high power density and fast response storage technologies (typically milliseconds) are preferred for microgrid autonomous operation [4].

4.2.4.2 Hydrogen Fuel Cell

Hydrogen fuel cell (HFC) converts chemical energy in fuel (i.e., hydrogen (H_2) and oxygen (O_2)) into electrical energy and its by-products are heat and water. Fuel cells consist of two electrodes (i.e., anode and cathode) separated by an electrolyte. The anode receives fuel (H_2) and then the oxidant (i.e., O_2) is fed into the cathode. Accordingly, the power is generated by passing ions from one electrode to the other [8]. In order to supply the microgrid with a certain amount of active and reactive power, the fuel supply has to be regulated to maintain the certain DC voltage to meet the inverters' input requirements. The main disadvantage of the fuel cell integrated into the microgrid is that its response time is variable, varying from milliseconds to minutes. Therefore, more research is needed to increase the response of fuel cells for microgrid applications [5]. Moreover, the overall efficiency of the fuel cells is relatively low compared with other storage technology. Interestingly, the HFC has the ability to storage a huge amount of energy for long time periods as hydrogen, which makes them very attractive for future microgrid applications [2].

4.2.4.3 Capacitive Energy Storage System

Capacitive energy storage system (CESS) plays an important role in controlling the microgrid frequency in the short term in case of a disturbance in the microgrid. In the event of a disruption in the microgrid, CESS can act as a dynamic frequency stabiliser for a short period of time until the DERs respond to the disruption [5].

Super-capacitors technology is one of the CESS types, which has an extremely high capacitance, capability for fast charging and discharging and very long cycle life. Therefore, super-capacitor technology can be charged and discharged at the same rate and very fast. Compared to BESS, CESS can be a better option for balancing the energy in the microgrid, especially in the power events that require short transients [2].

4.2.4.4 Flywheel Energy Storage System

Flywheel energy storage system (FESS) stores the energy as mechanical energy (i.e., rotational energy) by using the conservation of angular momentum principle. The FEES has the generalised concept of an accumulator. Accordingly, the energy can be storied by spinning the flywheel around its symmetry axis to save the electrical energy in a form of kinetic energy. This energy can be restored as a large amount of electrical power output and in a short time period (by slowing the spinning of the flywheel) once it is needed. Therefore, FESS can be suitable for the microgrid applications that required high amount of power in short time period with a high number of charging and discharging cycles [2].

4.2.4.5 Thermal Energy Storage System

Thermal energy storage systems store the energy in the form for of sensible or latent for several applications, including heating and colling as well as electrical generation. Thermal energy storage systems are used in several electrical grid applications in capacities of few kW to tens MWs. This technology is interesting in microgrids, which can use the access energy from DERs (e.g., rooftop PV system) in residential premises to heat up the water during low-demand periods [2].

4.2.5 Electric Vehicles Supply Equipment

Electric vehicles supply equipment (EVSE) provides a new load for the microgrid system, providing a great deal of flexibility. EVSE contains the electric vehicles (EVs) and the charging/discharging stations. Currently, the vehicle-to-grid (V2G) strategy can take advantage of the EVs load flexibility and distributed storage capacity of the vehicles when parked and linked to the grid. The power converters of EVs can inject power into the grid to support the grid system operation. One example of this support is participation in frequency regulation via a local controller capable of responding to frequency excursions or setpoints provided by a central frequency control system [13]. Based on their uptake and use patterns, EVs can help, hinder, complement, or constrain microgrids and related technologies. In addition, EVs can increase the total volume of electricity consumed and alter consumption patterns as well as flatten their load profiles.

The impressive growth trajectories forecasted for EVs, BESSs and digitisation present unprecedented challenges to grid network topologies and supervision, in terms of how EVSE are deployed, and the significant impact placed on energy systems. Interestingly, the coordinated bi-directional charging/discharging strategy can provide several ancillary services to the microgrid by improving the power quality, smoothing out peak demand and limiting the needs for new lines and transformer stations [14]. The main challenge of EVSE is forecasting the drivers' behaviours to model EVSE in the microgrid to ensure the high reliability of the operation, especially in islanded operation mode. The lack of regulations and standards in considering such technology in microgrid increases the challenge in the best representation of EVSE in microgrids. In addition, controlling the bi-directional charging/discharging processes may introduce some legal issues by limiting the customers to charge or discharge their EVs at any time period. As it is not easy to control the bi-directional charging/discharging processes at this stage, EVSE can be considered as a type of non-dispatchable DERs. Therefore, more research is needed to study the possible ways of representing the behaviours of the EVs owners/drivers as well as propose a proper business model to avoid any legal issues and encourage them to contribute to improving the reliability and stability of microgrids.

4.2.6 Power Converters

The bulk of microgrid components, specifically DC DERs, ESSs, and EVSE, require power electronic interfaces (e.g., DC/DC, AC/DC, DC/AC (i.e., inverters) or AC/DC/AC) to maintain the microgrid's voltage and frequency, and to convert DC power into grid-compatible AC power [4].

4.2.6.1 DC/DC Converters

DC/DC converter plays a major role in operational stability of the intermittent DERs, ESSs and EVSE in the microgrid to obtain a synchronised power delivery. In ESSs and EVSE, DC/DC converters are not just improving the charging/discharging efficiency, but they can also protect the ESSs from the under or over-charging issues [15]. Several topologies of DC/DC converters can be used in the microgrid based on the operation and the application. These topologies can be listed as follows:

- **Single-phase, non-isolated, bidirectional DC/DC converter** can be used to interface the energy storage systems with the DC bus. This topology is suitable when galvanic isolation and high-step-up or step-down of voltage is not needed.

Here, the bidirectional cascaded buck-boost converters can be used for the applications needed higher power and required energy sharing among distributed users based on load demand as they have a higher discharge current [16].
- **Dual-active-bridge DC/DC converter** is suitable for the applications requiring electrical energy at high frequency. This topology has many merits in ensuring electrical isolation, providing high power density, allowing bidirectional flow of energy, and offering an easy realisation of soft-switching control [17].
- **Bidirectional three-port DC/DC converter** works without the need for a power decoupling control, and it has a low circulating current. This converter is suitable for the applications that need high power density [18].

4.2.6.2 AC/DC Converters

AC/DC converters (also called rectifiers) are key important elements in the energy conversion applications, including their wide applications in microgrids. For instance, the AC/DC converters are used to charge batteries when they are connected to AC bus in the microgrids, to supply a DC load in the AC microgrids, to convert the generated AC power by wind turbines to DC power, and to regulate the imported AC power from the grid to DC or hybrid microgrids. Based on the operation principles and the desired applications, the AC/DC converters can be classified to [19, 20]:

- **Single- and three-phase AC/DC converters** rectify the AC input whether single- or three-phase to a DC output power. The three-phase AC/DC converters are preferred in systems with high power rating such as EVs charger technologies.
- **Half- and full wave AC/DC converters**. While the Half wave AC/DC converters required fewer components, the main advantage of full wave AC/DC converters is their high efficiency.
- **Uncontrolled (diodes-based) and controlled (silicon-controlled rectifiers (SCRs)/switches-based) rectifiers** are used to produce the desired DC output voltage level. Furthermore, based on the time that SCRs/switches are switched on/off, SCRs/switches-based rectifiers allow wide range control of the output voltage.

4.2.6.3 Inverters

The inverters (i.e., DC/AC converters) are the most used type of power converters in microgrids. The inverters are converting the DC power to AC power with high currents and voltages. The inverter is designed to deliver a controlled amount of power (active and reactive) to the grid. The grid side inverters are providing several primary functions as follows [4]:

- Control of the active and reactive power flow with the grid.
- DC-link voltage and synchronisation between the micro-sources and the grid.
- Voltage and frequency regulation.
- Harmonic filtering and unbalanced compensation.

- Initiate or support the system restoration.

The inverters are mainly categorised based on their operation, control in the grid as well as their grid supporting capabilities into grid-following and grid-forming inverters. The inverter control functions are an important concern in the design and operation of the microgrid.

Grid-following inverters

Grid-following inverters aim to synchronise the DERs with the grid voltage waveform. They adjust their output to track the current/power reference commands. Here, grid-following inverters are controlled using several strategies, including the well-known active and reactive control (PQ) strategy. The PQ-controlled inverter cannot impose voltage or frequency. Micro-source inverters (both controllable and non-controllable) are usually operated as grid-following inverters for both grid-connected and islanded operation modes [4]. The grid-following inverters require a voltage reference from other generators to operate. Accordingly, the inverters will shut down if they lose the voltage/frequency source. Here, grid-following inverters can support the microgrids autonomously by adjusting the output power as a function of the local voltage and frequency measurements. However, grid-following inverters have a limited response speed, which can potentially exacerbate disturbances [21].

Grid-forming inverters

Grid-forming inverters set their own internal voltage waveform reference and synchronise the DERs with the grid. Therefore, grid-forming inverters do not rely on external grid voltage, which allows them to operate independently of other generation. Accordingly, they establish the phase angle and magnitude of an output voltage. They are usually referred to as voltage source inverters (VSI) and are also capable of providing voltage and frequency regulation, especially during critical operation periods (e.g., emergency). Accordingly, grid-forming inverters can inherently help the stability of the microgrids by adjusting their active and reactive output powers instantaneously to meet the local voltage and frequency [21].

In Australia, AEMO is exploring the advance inverter technologies and has taken a gradual (i.e., stepwise) approach to obtaining various services from advanced inverters. These advance inverter technologies include grid-forming inverters to increase the capability from lowest to highest, which includes the capabilities of: (i) connecting inverter-based resources in weak grids; (ii) supporting system security; (iii) islanding operation; and (iv) system restoration/restart [21]. In the USA, the Hawaiian Electric Company (HECO) requests the new projects in 2021–2022 to include both standalone storage and hybrid PV-storage installations to be able to support and continue to operate the network (or microgrid) in a stable and reliable manner without the presence of synchronous generation resources (i.e., loss of grid reference) [22].

Smart inverters control functions

There are various smart inverter control functions that can increase the reliability, efficiency, and flexibility of the microgrid as well as increase its hosting capacity [23]. These smart inverter functions are supported by several standards and regulations, such as Institute of Electrical and Electronics Engineers (IEEE) Std. 1547-2018 [24] and AS/NZS 4777.2:2020 [25]. Accordingly, the smart inverter control functions are described as follows [23, 26]:

- **Constant power factor function**. The inverters can ensure a constant power factor of the system at any time period. This means that the ratio between reactive and active power should be proportional to each other. In order to achieve a constant power factor, proportionally low reactive power is injected or absorbed by the inverter to achieve a leading or lagging constant power factor, respectively. This will introduce some active power curtailment, which reduces the customers' revenues.
- **Volt-watt function**. The aim of the volt-watt control function is to manage the active power injection capability of the inverter interfaced with DERs according to mitigate any over-voltage issue in the microgrid in order to keep the voltage within the pre-defined limits. This feature cannot mitigate the low-voltage issues that may be introduced by single-phase DERs due to the mutual impedance between the phases (i.e., increasing the phases in one phase can reduce the voltage on the other phase and vice versa). Therefore, high under-voltage issues can be appeared at high DERs penetration levels using volt-watt only. Volt-watt can lead to high active power curtailment, especially at high DERs penetration levels, which significantly reduces the customers' revenues.
- **Volt-var function**. The volt-var control function aims to mitigate both over- and under-voltage issues by absorbing the reactive power to reduce the voltage or injecting reactive power to increase the voltage in the microgrids. Volt-var function can cause less active power curtailment, which increases the customers revenues compared with using volt-watt function only. However, volt-var control function can cause some power quality issues (e.g., poor power factor) and increase the magnitude of flowing currents in the conductors once the inverters are absorbing the reactive power from the microgrid, which may lead to overload the cables and transformers and cause an over-heating issue.
- **Volt-watt with volt-var function**. Regardless of the DERs penetration levels, the combination of the volt-watt and volt-car functions can play a significant role in keeping the voltage in the network within the maximum and minimum limits. In addition, this combination can reduce the active power curtailment significantly compared with volt-watt function. Moreover, it can reduce the larger amount of current flow from the substation to the transformer as it reduces the amount of the absorbed reactive power compared with volt-var function.

4.2.7 Other Components

In addition to AC and DC lines, power converters and power transformers, the electricity network components may also contain reactive power (i.e., VAR) support devices (e.g., capacitor banks, distribution system with distribution static compensator (DSTATCOM) and distributed static var compensation (DSVC)), electromechanical circuit breakers, solid-state switches, and AC/DC filters. Microgrids must also rely on a variety of control hardware and software connected over wireless or fixed wired communications networks in order to ensure a secure, coordinated, and efficient operation [4].

Summary

Microgrid is a combination of loads (e.g., sensitive/critical, non-sensitive/non-critical controllable and/or non-controllable loads), DERs (e.g., dispatchable, and non-dispatchable DERs) and power converters (e.g., inverters) with monitoring, control, and protection devices. A wide range of DERs and energy storage technologies with different characteristics can be used in microgrid applications. The selection of DERs and energy storage technologies mix depends on the potential of these technologies, capital and operation costs, size, charging/discharging time and efficiency, cycle life, power and energy densities, and environmental factors. Based on the reviewed national and international projects, ESSs, which are considered as dispatchable DERs, are one of the key components in microgrid projects, especially in islanded operation mode, to ensure meeting the demand. Most of the national and international projects are using lead-acid BESS technology and flow batteries. In addition, PV systems are the most used non-dispatchable DERs in microgrid projects around the county and worldwide. However, the selection of DERs mix depends on the potential of these DERs, costs, and environmental factors.

It is common in today's microgrid design to have one large grid-forming plant serve the entire microgrid. The grid-forming plant should consider dispatchable DERs (e.g., diesel generators, small-scale hydropower plant and ESSs). In this scenario, the single grid forming plant is responsible for maintaining power quality of the system (with the help of other voltage regulating devices, if available) and its performance requirements should be specified to achieve this goal. Furthermore, most of the used inverters in microgrid projects are the grid-following inverters with some providing grid-supporting functionality. In addition, it is worthy to mention that an active network management strategy based on the smart inverter functions can be defined to ensure meeting the voltage and thermal constraints in the microgrid. Therefore, they are often implicitly or explicitly determining the setpoint control for DERs by determining the exact value of the DERs exports that will best allow some operational objectives to be achieved. However, there are many different types and implementations of grid-forming inverters control systems, with trials underway in Australia and worldwide to demonstrate their grid supporting capabilities. Accordingly, AEMO (Australia) and HECO (USA) recommend prioritising the use of grid-forming capability on energy storage elements such as batteries both due to a large number of energy storage projects in the interconnection queue along with the availability of an energy buffer.

4.3 Modelling Techniques

Several techniques can be deployed to model the performance and behaviour of the microgrid components. These modelling techniques are mainly based on two main approaches either a model-based approach, or a data-driven approach. The model-based approach includes the explicit electrical models, which are suitable for steady-state and dynamic simulations a well as can allow the "what if" scenarios. In contrast, the data-driven approach needs a high volume of data, but it is challenging to support "what if" scenarios and dynamic simulation [27]. Data-driven approach can be conducted using machine and deep learning algorithms.

4.3.1 Modelling of Electrical Networks

Usually, the electricity networks are represented by a "node and branch" model, where the nodes are the connection points where the customers can obtain their electricity supply or connect their DERs to, and the branches represent the physical assets in the network (e.g., lines, protection equipment, and transformers). The connection points are usually identified by a national metering identifier (NMI) and associated with a meter (e.g., basic, or advanced) for market operation and participation purposes.

4.3.1.1 Model-Based Approach

The model-based approach [27] contains explicit electrical models of the corresponding components in the distribution/transmission network (i.e., bus, branch, loads, generators, DERs, shunts, etc.), including the network topology and parameters. The explicit electrical models should be characterised to allow running power flow simulations using a tool for different demand/generation scenarios, or "what-if" scenarios. This approach is the most used modelling approach by DNSPs in Australia and worldwide for designing, planning, and improving the distribution networks, including microgrids. The model-based approach can be considered as a highly effective approach for investigating the effectiveness of various solutions or scenarios. In a model-based approach, the planning standards and guidelines for a distribution network should be defined. Then, power flow can be conducted by including the demand/DERs scenarios of interest. Several considerations can be followed by DNSPs to ensure the best practices in implementing and providing better quantifications and insights, including low-voltage (LV), medium voltage (MV) and high voltage (HV) feeders, as follows:

- Selecting the LV/MV feeders is carried out as a trade-off between the quality and availability of data, existing DERs, and the complexity of extracting the corresponding data.
- Integrating HV-MV-LV feeder models with LV detailed feeders down to the single-phase household level.
- The DERs should be modelled, including the specific settings/requirements, if exists (i.e., smart inverter control functions, export limits, etc.), to adequately capture their interactions with voltages and power flows.
- The uncertainty should be considered, which includes the future locations and sizes of DERs, the meteorological data, and household demands.

The advantages of this approach are: (i) Its network elements and participants can be added, removed, or changed; (ii) It supports the "what-if" scenarios; and (ii) It can be used to deploy several control actions and tested through simulations. However, its main disadvantage of this approach is the model simplifications arising from lack of critical visibility or network information, which can lead to grossly incorrect results.

4.3.1.2 Data-Driven Approach

The data-driven approach could be a very promising alternative to model-based approaches that can save both time and efforts [28]. Unlike the model-based approach, the data-driven approach does not require explicit electrical models, rather it uses historical data for the system status and component behaviours. These data could be collected using real-time measurements and/or simulation data for huge types of scenarios supported by internal and external validations. The accuracy of this approach strongly depends on the volume and resolution as well as the accuracy of the collected data. Several techniques can be used to map and clean the data as well as find the missing data points to increase the accuracy of this approach.

The strong linear correlation between the aggregated customer consumption in each node/transformer/feeder/network associated with the voltage level at a given time can enable DNSPs to apply some optimisations related to several scenarios using the historical data. A regression machine learning model (i.e., supervised univariate regression model) is used in the "Advanced planning of PV-Rich Distribution Networks" project by a research team at the University of Melbourne [28] to estimate the hosting capacity at several LV feeders in Victoria. The developed model required a huge amount of smart meter data to secure an acceptable range of accuracy. To ensure the realistic implementation of the data-driven approached by providing better quantifications to increase the reliability of the results, several considerations should be followed by the DNSPs as follows:

- The historical data of all customers' consumption and voltage levels can be extracted from smart meter data in a given LV network. Therefore, following this approach requires high visibility of the network and customers' data (i.e., high penetration of smart meters).
- The missing and inconsistent values can be cleaned from the database. Therefore, this approach is not straight forward and requires some pre-processing stages.
- The obtained results from this approach should be internal validated by running trials of the smart-meter-driven approach.
- It is recommended to reinforce the training data pool with simulation data for some of the unexpected, unique, or unusual scenarios (e.g., tap changing in the zone substation's OLTC actions).

The advantages of this approach are: (i) it does not require the explicit electrical models of the network components or continuous power flow simulations; and (ii) it can save a lot of resources and time for new DER connection requests. While this approach is limited in supporting the planning and expanding the networks or any application requires changing the fundamental/configuration of the network as it does not support the "what-if" scenarios. In addition, this approach needs a high penetration level of smart meters and data servers supported with data cleaning and storing tools.

4.3.2 Modelling of Electrical Loads

Electrical loads are considered as power consumption units. The nature of the loads depends on several factors, including load factor, power factor, demand factor and a utilisation factor of the system. In general, the electrical loads can be represented in form of voltage, current and power factor; active power and power factor, active power, and reactive power; and complex power and power factor. The load consumption can be modelled as a constant load, constant load associated with probability function or time-series actual data. The constant loads can be categorised based on the representations of the voltage magnitudes, such as the constant active and reactive power model, constant impedance model, constant current model, ZIP load model (i.e., a combination of impedance, current, and active and reactive models) or any other combination of them. An accurate load model should be considered time-series actual demand data for an individual customer instead of a group of customers. The voltage of the load is always the voltage at the node, or the LV feeder based on the location of the connection. Therefore, the accurate load model should allow the change of the demand over time with respect to the supplied voltage, and be able to characterise the current at the customer side [23].

4.3.3 Modelling of Distribution Energy Resources

DERs are the main energy sources in microgrids, especially in islanded operation mode. Therefore, the accurate modelling of DERs performance strongly affects the accuracy of the design results is selecting the type as well as the optimal location and size. Moreover, it also strongly affects the operation of the microgrid and business models. The uncertainties should be considered in modelling DERs (e.g., meteorological and weather variables) as the high fluctuation of the output power can cause severe problems if it does not take into consideration. The selection of the proper model depends on the application as well as the trade-off between the accuracy, simplicity, and simulation burden. Accordingly, the accuracy of the results depends on the model representation and the availability and accuracy of the data. In general, the above-mentioned approaches in modelling the network are both options to represent the characteristics and the performances of the DERs.

4.3.3.1 Modelling of Small-Scale Hydropower Plants

The output power from the small-scale hydropower plants is a function of the water discharge, hydraulic head, water density, acceleration due to the gravity, the efficiency of the turbine and the generator [29]. The power produced has a direct relationship with amount of discharge and head, which increases power produced with the increasing amount of discharge and head [30]. The isolated asynchronous/induction generators (e.g., fixed-speed induction generators (FSIGs) with squirrel cage rotors, and doubly-fed induction generators (DFIGs)) can be used to model the small-scale generators. The generator can be modelled as a power source that injects the active and reactive power to the network according to the load consumption, losses, and injected power from the other resources. Accordingly, the generator in steady-state can be modelled as operating in PQ mode or PV mode. While a detailed generator model is needed for dynamic (i.e., stability) simulation.

In the hydroelectric power generation system, the kinetic energy of the flowing water is converted to mechanical energy using water turbine which then drives the generator to produce electricity. The energy produced by the turbine relies on the fast flow velocity of water, weight per unit volume, specific gravity, and hydraulic head. The power output of the hydropower plant is proportional to the amount of water discharge and head. For any hydroelectric generation system, the measured value of accessible power can be calculated using the following equation [29].

$$\boldsymbol{P} = \eta\rho g\boldsymbol{HQ}, \tag{4.1}$$

where P denotes the produced electrical power in Watts, ρ presents the density of water (1 kg/L), η is the turbine and generator efficiency (between 1 to 100%), g is the acceleration due to gravity (9.81 m/s^2), H is the hydraulic head and Q is the discharge in m^3/s.

4.3.3.2 Modelling of Reciprocating Internal Combustion Engines

The reciprocating internal combustion engines can represent the generators that are operated using natural gas, diesel oil, petroleum, and biofuels. These generators are usually modelled as AC synchronous generators (e.g., permanent magnet synchronous generators (PMSGs) and electrically excited synchronous generators (EESGs)). The mathematical model can be represented in steady-state simulation as a power source operating either in P-Q mode or in P-V mode with a cost function [31]. As with the other generators, detailed dynamic model is required to study the stability in microgrids, including the speed control, temperature control, fuel control system and compressor turbine [8].

The operating cost of the diesel generator depends primarily on the cost of the fuel. The cost function of a diesel generator is given as [32]:

$$\boldsymbol{C_{active}(t) = a_c P_{de}(t)^2 + b_c P_{de}(t) + c_c}, \tag{4.2}$$

where $C_{active}(t)$ and $P_{de}(t)$ denotes the generation cost and active power output from diesel generator at any time, respectively, and a_c, b_c, and c_c are the coefficients of the generation cost function.

The above cost function provides the cost of active power generation only. The total cost of providing reactive power support can be presented as follows [31]:

$$\boldsymbol{C_{reactive}(t) = Aq + B}, \tag{4.3}$$

where $C_{reactive}(t)$ is the total cost of reactive power, q is the rating of generator in MVAR, and A and B are the variable cost and fixed cost, respectively.

Alternatively, the total cost of reactive power support can be expressed in the form of the total cost of reactive power needed to bring the system minimum voltage in a certain level. This can be realised using the convex hull pricing model, which is a well-behaved convex approximation of the total cost function. The relationship between the approximated total cost of reactive power and the system minimum voltage can be represented as follows [31]:

$$\boldsymbol{C^{CH}_{reactive}(\mathrm{t}) = 643{,}600\mathrm{V}^2_{min} - 1{,}056{,}007V_{min} + 427{,}786}. \tag{4.4}$$

4.3.3.3 Modelling of Microturbine Systems

A typical microturbine model was developed by connecting the gas turbine to PMSG. The steady-state representation of the microturbine models is the same as that discussed in small-scale hydropower plants and small reciprocating internal combustion engines. However, several models are used in the literature to characterise the dynamic behaviour of the microturbines, including a heavy-duty gas turbine model

[33]. An example of a dynamic model of microturbine systems with actual field data and various types of gas is developed in [8] and in [34].

4.3.3.4 Modelling of Solar Photovoltaic Systems

PV systems are a combination of PV modules and power converters. Here, the performance of PV modules is discussed, while the modelling of power converters is discussed later in this subsection. The PV modules consist of several PV cells connected in series and parallel. The modelling of PV modules could be done either by modelling the whole PV module or modelling the PV cells and then find the cumulative performance of the PV module based on the number of PV cells in series and parallel with considering the shading effects. The performance of PV cells/modules is strongly dependant on meteorological variable, especially solar radiation, and ambient temperature. Therefore, the non-linear relationship between these variables and PV performance should be accurately represented. Several techniques are deployed to characterise the performance of PV modules, including linear, regression, and machine and deep learning models. Linear models have several limitations in characterising the non-linear behaviour of PV modules, and introduces several assumption and simplifications, which affect the accuracy of the model. While regression models can be useful to model PV modules, but it could not handle the high uncertainties. Machine and deep learning models can overcome the aforementioned limitations, but they need high volume of data with high resolutions. However, implicit exponential equations associated with diode and solar cells can be powerful model to express the behaviour of PV cells/modules if their parameters are extracted correctly based on the provided electrical specifications and meteorological variables. These parameters can be extracted using numerical, analytical or artificial intelligence (AI) methods [23]. The implicit exponential equations are resulted in providing the output current, voltage and power considering the temperature and solar radiation effects, which are important to control the PV systems to extract their maximum power under variable operating conditions [35].

The model describes the current-voltage (I-V) and power-voltage (P-V) characteristics over a wide range of temperatures and solar radiation values. The single-diode PV cell model, known as five-parameters model, has been one of the most widely used model for characterising performance of PV modules. Details of the five-parameters model can be found in [36]. To get the I-V characteristic curve of the PV module for any value of solar irradiation (G) and temperature (T) a generalised model is required. Chenlo's model is one of the popular simplified and generalised models used for this purpose. The translation procedure of Chenlo's model to find the new I-V pairs is visualised in Fig. 4.3.

In this model, the short-circuit current ($I^{PV,sc,ref}$) and the open-circuit voltage ($V^{PV,oc,ref}$) of the PV module can be collected from the datasheet of the PV module for a particular solar radiation (G^{ref}) and temperature (T^{ref}) at the standard test condition (STC). Based on that, the new $I_k^{PV,sc}$ and $V_k^{PV,oc}$ can be determined for the new values of solar radiation (G_k) and temperature (T_k) at each time step as follows

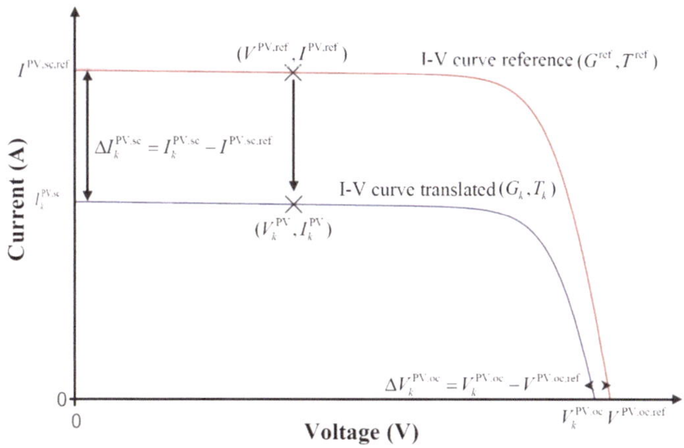

Fig. 4.3 Chenlo's model to find the new I-V pairs at any G and T values

[36]:

$$I_k^{PV,sc} = I^{PV,sc,ref} \cdot \frac{G_k}{G^{ref}} + \alpha \cdot (T_k - T^{ref}), \tag{4.5}$$

$$V_k^{PV,oc} = V^{PV,oc,ref} + V_k^{th} \cdot ln\left(\frac{G_k}{G^{ref}}\right) + \beta \cdot (T_k - T^{ref}), \tag{4.6}$$

where V_k^{th} stands for the diode thermal voltage, which can be calculated based on the cell temperature as a function of ambient temperature and time, α is the temperature coefficient of the short-circuit current, and β represents the temperature coefficient of the open-circuit voltage.

The deference between the new calculated $I_k^{PV,sc}$ and the $I^{PV,sc,ref}$ ($\Delta I_k^{PV,sc}$) as well the deference between the new calculated $V_k^{PV,oc}$ and the $V^{PV,oc,ref}$ ($\Delta V_k^{PV,oc}$), respectively, is given as follows [36]:

$$\Delta I_k^{PV,sc} = I_k^{PV,sc} - I^{PV,sc,ref}, \tag{4.7}$$

$$\Delta V_k^{PV,oc} = V_k^{PV,oc} - V^{PV,oc,ref}. \tag{4.8}$$

Utilising the known I-V parameters (i.e., $I^{PV,sc,ref}$ and $V^{PV,oc,ref}$) and calculated values from the above equations, the new I-V pairs (i.e., $I_k^{PV,mpp}$ and $V_k^{PV,mpp}$) at maximum power point based on the new G and T at time k can be determined as follows [36]:

$$I_k^{PV,mpp} = I^{PV,mpp,ref} + \Delta I_k^{PV,sc}, \tag{4.9}$$

$$\boldsymbol{V}_k^{PV,mpp} = \boldsymbol{V}^{PV,mpp,ref} + \Delta \boldsymbol{V}_k^{PV,oc}. \tag{4.10}$$

4.3.3.5 Modelling of Wind Turbine Systems

The wind turbines can be represented using a wind power curve, which is mostly provided by the manufacturer. The wind power curve could be provided as a figure or equation. The wind power curve is mostly divided into three main regions: a "no wind power" region (where the wind speed is below the cut-in speed), a "de-rated" region, and a "rated power region". The wind turbine will shut down once the wind speed is higher than the cut-off speed to protect the blades and the generator. The output power from the wind turbine is a function of air density, swept area of the turbine, nominal wind velocity, and the nominal power coefficient of the whole wind generator [37]. The power produced has a direct relationship with wind velocity, which increases the power produced with increasing the wind velocity. The output power from the wind turbine varies based on the generator type, which can be, DC, AC, or AC with variable frequency. In general, a power converter is required to integrate the wind turbines into the microgrid to ensure the synchronisation requirement and for control actions [38]. The type of generator could be DC generators (e.g., shunt-wound DC generator), AC synchronous generator (e.g., PMSGs and EESGs), AC asynchronous/ induction generators (e.g., fixed-speed induction generators (FSIGs) with squirrel cage rotors, and doubly-fed induction generators (DFIGs)) or switched reluctance wind turbine generator. For steady-state simulation, the generator in wind turbine can be modelled as in small-scale hydropower plants either based on its operating in PQ mode or PV mode. However, the detailed generator model is needed for dynamic (stability) simulation [39].

The electricity generated from the wind energy systems entirely depends on the wind speed. The wind power conversion model can be represented [40] as:

$$\boldsymbol{P}_W(t) = \begin{cases} \boldsymbol{0}, & \boldsymbol{v}(t)\langle \boldsymbol{v}_{in} \text{ or } \boldsymbol{v}(t)\rangle \boldsymbol{v}_{out} \\ \frac{1}{2}\pi \boldsymbol{r}^2 \rho \boldsymbol{v}(t)^3, & \boldsymbol{v}_{in} \leq \boldsymbol{v}(t) \leq \boldsymbol{v}_{rate} \\ \boldsymbol{P}_W^{rate}, & \boldsymbol{v}_{rate} \leq \boldsymbol{v}(t) \leq \boldsymbol{v}_{out} \end{cases}, \tag{4.11}$$

where $P_W(t)$ is the power generated at time t from the wind turbine of rated power of P_W^{rate} (kW) having rotor radius (r) in m, $v(t)$ is the measured wind speed at time t in m/s, v_{in} is the cut-in wind speed in m/s, v_{out} is the cut-out wind speed in m/s, v_{rate} is the rated wind speed in m/s, and ρ is the air density in kg/m^3.

4.3.4 Modelling of Storage Systems

The models of the BESS and HFC technologies are just mentioned here as the BESSs is one of the most used technologies in microgrid projects and HFC is a promising technology.

4.3.4.1 Modelling of Battery Energy Storage Systems

The BESSs are modelled using several representations, including the Thevenin-equivalent mathematical model [41], kinetic battery model [42] (represents the sensitivity of storage capacity to discharge rate), and detailed model consisting of a controlled voltage source and an internal resistance with current discharge characteristics [8]. In general, the BESSs can be modelled as a constant power load during the charging period, and generator than can inject the active power to the network during the discharging period. The BESSs model is always subjected to its power rating and it stored energy capacity [43]. The status of the BESSs is usually modelled based on the SOC (the ratio of the available capacity and the maximum possible charge that can be stored in the BESS). The way of modelling the SOC can be varied based on the application and availability of the data. In microgrids, the BESSs are interfaced with several power converter topologies to convert the energy to the right type to be used and to ensure a high charging/discharging efficiency. The control of the BESSs is required to ensure the protection of BESSs during the charging and discharging as well as to support the balancing with the demand-side in the islanded operation mode (i.e., dispatching the BESSs to follow the load) [44].

The energy storage system is modelled as a constant power load during charging and as a generator during discharging [43]. The charging and discharging operation are always subjected to its stored energy capacity and power rating. The general model of the energy storage system is illustrated in Fig. 4.4.

The storage system is operated in three modes: charging, discharging, and idling. The net increase/decrease of the stored energy in the ideal storage unit during charging/discharging operation are subjected to the different incurred losses (i.e., inverter, idling and charging/discharging losses). The variation of energy stored (E)

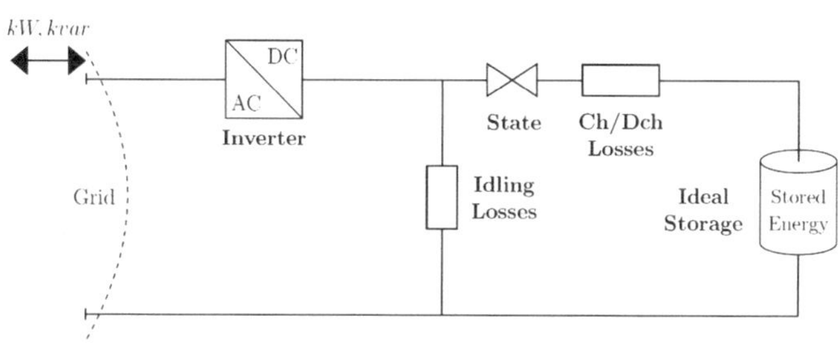

Fig. 4.4 General model of the energy storage system

in the storage system at any time is dependent on its previous state as presented in the following equation [43].

$$E(t+\Delta t)=\begin{cases} E(t)+P_{eff}(t)\Delta t, & \textbf{charging mode} \\ E(t)-P_{eff}(t)\Delta t, & \textbf{discharging mode} \end{cases}, \tag{4.12}$$

where $P_{eff}(t)$ can be obtained as [43]:

$$P_{eff}(t)=\begin{cases} P_{in}(t)-P_{losses}^{ch}(t), & \textbf{charging mode} \\ P_{out}(t)+P_{losses}^{ch}(t), & \textbf{discharging mode} \end{cases}, \tag{4.13}$$

where P_{in}/P_{out} are the power at the storage unit's grid interface terminal during charging/discharging operation.

4.3.4.2 Modelling of Hydrogen Fuel Cells

The power generation from the hydrogen fuel cells (HFCs) depends on several factors, including the oxygen usage rate, hydrogen usage, energy content of hydrogen, air inlet flow rate, air exit flow rate, rate of water production, and heat production. Several models of the HFCs dynamics and its power generation are proposed and validated in the literature, such as the mathematical dynamic electrochemical model of the proton-exchange membrane fuel cell (PEMFC) [45], control architecture of PEM fuel cells [46], and optimisation-based model for identifying unknown variables in a solid oxide (SO) fuel cell [47].

4.3.5 Modelling of Electric Vehicles Supply Equipment

The electric vehicles supply equipment (EVSE) consists of the EVs and charging/discharging stations. The modelling of the EVs varies based on the application and the use case(s). Several modelling approaches are considered in the literature to model the EVSE. Here, the EVs can be modelled as a moving BESSs. However, the charging and discarding rates and time depend on several factors, such as the behaviour of the EVs owners/drivers, the driving rates and distances, and energy prices [48]. The V2G strategy introduces several limitations in modelling the EVSE as the behaviour of the EVs owners/drivers is not easy to be expected. Therefore, an accurate model should be considered to express the behaviour of EVSE in microgrids, especially during the islanded mode. In most cases, the microgrid operator should have the authority to control the charging and discharging periods and rates, which may introduce charging/discharging curtailment penalty [13]. In this context, the EVSE can be simply modelled as a load, where this model is not realistic as it does not consider V2G and the microgrid operator cannot control it. However, some projects modelled the EVSE as a controllable load. This assumption allows the microgrid operator to control the charging rates and periods, but it is neglected the V2G strategy. To overcome that limitation of the previous models with including V2G strategy, the EVSE is modelled as an intermittent ESSs, where it considered the V2G strategy, but without any control on the charging/discharging rates and periods by the microgrid operator. Here, the more realistic model that can allow the microgrid operator to control the charging rates and periods and consider the V2G strategy is a controllable ESSs model. All of the mentioned EVSE models require an accurate model to express the behaviour of the EVs owners/drivers and reflected on the amount of power either injected to the grid or consumed from the grid. These concerns may affect negatively on the reliability of the microgrids. Thus, to avoid the low reliability issues in the microgrids with considering the EVSE, several technical requirements can be followed as [4]:

- Optimal location of EVSE and local generating resources to mitigate the distribution system overloads and bottlenecks.
- Optimal charging and discharging of EVs and ESSs fleet in order to reduce overall costs.
- Optimal capacities for PV, EVSE, generators, and CHP can lower the of the generating mix for EVs fast charging and truck fleets.
- Mitigate the lower reliability issues that can be introduced by the EVSE in microgrid during the islanded operation mode by controlling the charging and discharging processes.

EVs and charging/discharging stations are the main components of the EVSE. The modelling of EVs differs depending on the application and use case(s). Several modelling methodologies for the EVSE are addressed in the literature. Each of those modelling methodology has its advantage, disadvantage and limitations as follows:

- **Modelled as a load**. The EVSE can be modelled as a load. This model is the simplest model for implementation. However, it requires historical data. This model does not support vehicle-to-grid (V2G) strategy and does not allow the charging/discharging control.
- **Modelled as a controllable load**. The EVSE can be modelled as a controllable load, which provides the feature of controlling the EVs charging but no the discharging. Therefore, this model also does not support V2G strategy. Similar to the previous model, historical data are required to represent this model.
- **Modelled as an intermittent electrical storage battery**. The EVSE can be modelled as an electrical storage battery. This model can characterise the charging and discharging modes but without controlling them. Therefore, this model can characterise the V2G strategy but without any control.
- **Modelled as a controllable electrical storage battery**. The EVSE can be modelled as an electrical storage battery with providing the control feature for the charging and discharging modes. Therefore, this model can characterise the V2G strategy with control. Here, controlling the charging and discharging status

(import/export limit) could require certain curtailment limit, which should be pre-defined by the DNSP.

4.3.6 Modelling of Power Converters

Modelling of power converters is essential not only to understand their dynamic performance but also to predict any possible instability issues in the microgrids and, consequently, preserve stable operation of the microgrid as required. Several approaches can be used to model power converters, including:

- **State-space modelling**. This modelling technique is the most accurate modelling method. It can be used to provide a complete insight into the system performance (e.g., eigenvalues, bode plots, root locus, controllability, observability, and stability analysis) [49].
- **Impedance-based modelling**. In this method, the converter is represented by its equivalent Norton/Thevenin circuit. Then, this model can be used for author applications such as evaluation of the system stability based on impedance ratios according to the generalised Nyquist criterion [50].
- **Black-box modelling**. This modelling technique is suitable for large-scale system with many converters as it is very difficult to derive the system state-space modelling. Black-box modelling relies on identifying the system transfer function from the measured frequency response [51].

4.4 Data Requirements for Modelling a Microgrid

There is a need to validate the existing data and gather new, previously untracked data based on the significant use case of designing and operating the microgrids. Multiple source systems like geographic information systems (GISs), customer information systems (CISs), asset management systems, and engineering design tools are involved in data collection and gathering processes. A wide range of techniques is employed, including correlating information from disparate repositories, searching paper records, leveraging validation algorithms, and performing extensive field surveys. After gathering and validating data, one of the critical challenges is to convert data to the planning tool format while ensuring consistency and accuracy of data. A certain level of manual support is needed to complete this process. Achieving a microgrid model with the requisite level of data integrity and completeness is often a multi-year, multi-tens-of-million-dollar process. Therefore, the type and level of data can be determined based on the use case(s) of microgrids.

A large amount of validated data is required to achieve an accurate power flow result. Missing and/or inaccurate network data/information is one of the significant concerns of the current utilities. Therefore, modelling accurate network models requires a higher degree of grid model accuracy in connectivity information and electrical parameter data. Besides, rectifying data quality and completeness issues can be highly time-consuming and expensive for DNSPs. Capturing the needed sets of information having 100's to 1000's customers, devices, and characteristics that are dynamic in time is not trivial in distribution systems. The data requirements to model the components in a microgrid are listed but not limited in Table 4.1.

4.5 Power Flow

Power flow aims to obtain the voltage magnitudes and angles at all nodes, and the currents flow in all branches by giving the topology and components of the network/feeder and set points of connected loads and DERs. The microgrids are mostly contains three-phases loads and DERs; therefore, unbalanced power flow analysis is required in microgrids to ensure the representation of the phase imbalances. In unbalanced power flow, the variables can be expressed using non-linear equations either in terms of voltage phasors, as in bus injection model (BIM) or in terms of squared voltage magnitudes and squares of the current flow as in the branch flow model (BFM). Several algorithms can be used to solve the power flow equations in microgrids, which are used by several power flows tools, as follows [23]:

- Newtown-iteration algorithm.
- Backward-forward sweep algorithm.
- Current injection algorithm.
- Holomorphic embedding algorithm.

The key to solving the unbalanced PF equations is determining the three-phase voltages at buses and the three-phase currents through all the conducting components. It is essential to consider both the topology of the network and its components, as well as the set points of the connected loads and generators, in order to obtain the solution of the non-linear unbalanced PF equations [23]. It is possible to use AC and/or DC power flow in combination for simulation and optimisation (e.g., optimal power flow (OPF)). Optimising is the process of finding the best solution to a problem. It is not always necessary to start the optimisation process with an accurate model. As a starting point, the inaccurate model can be utilised, try to optimise it, and plug the solution into the accurate model for reiterate. As a result, a good enough solution could be achieved in a fraction of the time.

4.5.1 Steady-State Power Flow

The steady-state power flow can be assessed based on two main analysis frameworks. The first framework is a snapshot stochastic analysis. This framework aims to determine the 'worst case' feeder response, which would be occurred for an intermediate change from zero to full DER output. The second framework is a time-series analysis

Table 4.1 Data requirements to model the components in a microgrid

Microgrid component		Data required
Network/feeder		• Generators' locations and specifications • Line's types, lengths and specifications • Transformer sizes, specifications, and connection points associated with the phases • Transformer adjustment data (e.g., minimum, and maximum taps and voltages) • Protection devices specifications • Network regulations and standards • Size, location, number of phases associated with the connected phase, and generation profile of any existed DER, ESSs and EVSE
Load		• Zone sub-station load • Participant segment (residential, commercial, farm) • Connection information (feeder and location) • Customer demand, including average load, peak load, total energy usage (e.g., peak, and off-peak) • Time-series active and reactive power; active power and power factor; active or reactive power and complex power; voltage, current and power factor • The number of phases associated with the connected phase
DERs	Small-scale hydropower plants	• Electrical specifications of the generator and turbine • The hydraulic head (i.e., the difference between the upstream and downstream water level) • Time-series data for the low-rate history or the water velocity history or the water discharge rate
	Reciprocating internal combustion engines	• Generator type, specifications, and limits • Fuel type and flow rate • Fuel cost function
	Microturbine	• Generator type, specifications, and limits • Fuel type and flow rate • Fuel cost function
	PV module	• Electrical specifications, including current-voltage or power-voltage curves • Time-series of the metrological data, especially solar radiation, and ambient temperature • Tilt and azimuth angles
	Wind turbine	• Electrical specifications of the generator and turbine, including the power curve • Mechanical specifications, including the height of the turbine, swept area of the turbine, and nominal power coefficient of the whole wind generator • Time-series of the metrological data, especially wind speed and wind direction
Storage systems	Battery energy storage system	• Electrical specifications

(continued)

Table 4.1 (continued)

Microgrid component		Data required
	Hydrogen fuel cells	• Fuel cell specifications • Oxygen and hydrogen usage rate • Air inlet flow rate and air exit flow rate • Rate of water production, and heat production
EVSE		• Electrical specifications of the EVs • Location and capacity of charging/discharging stations • Behaviour of the EVs owners/drivers • Time-series data of the charging and discharging history
Power converters		• Electrical specifications • Active power control settings (based on the standards)

which can be considered as a snapshot stochastic analysis with time varies based on time-series demand/generation data. It also can be carried out using specific DER deployments selected from the snapshot analysis results. The time-series analysis reflects more "realistic" feeder response results, which would be occurred according to measured load and meteorological data [52].

The power flow equations in an electrical power system aim to represent the relationship between the voltage phasors and the power injected into a node via units and shunts. Power flow equations are essential for conducting analysis and expressing the operation of an electrical power system in many optimisation and control problems.

The bus injection model (BIM) is a representation of the electrical quantities at each bus, is followed in the implementation. Here, the BIM is modelled using the I-V and voltage-based formulations. The I-V formulation relates the voltage, current injection, and power injection variables, where all current variables are omitted from the BIM representation, resulting in reactive and active power flows being represented solely as a function of the voltage differences between connected buses. Accordingly, the series impedance is consequently used in admittance form, which makes it improbable to represent zero-impedance branches [23].

The BIM model can be represented based on Ohm's law as follows:

$$\boldsymbol{Y}_{t\psi} \times \boldsymbol{V}_{t\psi} = \boldsymbol{I}_{t\psi}, \tag{4.14}$$

where $Y_{t\psi}$ stands for nodal voltage vector, $\mathrm{V}_{t\psi}$ is the nodal voltage vector, $I_{t\psi}$ represents the nodal current injection vector, ψ is the phase and t is the time.

Accordingly, the nodal complex power injection vector ($S_{t\psi}$) can be defined as:

$$\boldsymbol{S}_{t\psi} = \boldsymbol{V}_{t\psi} \cdot \left(\boldsymbol{I}_{t\psi}\right)^{H}. \tag{4.15}$$

In this context, the variable complex power flow in the series element in π-section ($s_{mat\psi}$) can be represented as:

$$s_{mat\psi} = v_{mat\psi} \cdot \left(i_{mat\psi}\right)^{*}, \quad (4.16)$$

$$s_{mat\psi} = P^{NET}_{mat\psi} + jQ^{NET}_{mat\psi}, \quad (4.17)$$

where $v_{mat\psi}$ is the nodal complex power injection vector element, $i_{mat\psi}$ stands for the nodal current injection vector element, $P^{NET}_{mat\psi}$ is the net active power vector element, and $Q^{NET}_{mat\psi}$ represents the net reactive power vector element.

Here, the current flow between bus m and n in π-section model ($f_{mnat\psi}$) can be mathematically represented as:

$$f_{mnat\psi} = \left(v_{mat\psi} - v_{nat\psi}\right) \cdot y_{mnat\psi}, \quad (4.18)$$

$$-F_{mn} \le f_{mnat\psi} \le F_{mn}, \quad (4.19)$$

where $y_{mnat\psi}$ is the nodal admittance matrix element, and F_{mn} stands for the current flow capacity (i.e., ampacity).

The current injection algorithm is used to solve the power flow equations based on a current injection method. This representation is expressed in a combination of rectangular and polar coordinates [23]. In this work, a PQ bus is represented by two equations consisting of the imaginary and real components of the current injection mismatches, which are expressed by means of the rectangular voltage coordinates.

4.5.1.1 Snapshot Analysis

In general, the snapshot analysis can be conducted for two base load levels: the absolute maximum and absolute minimum. These load levels are utilised to derive a bounding worst-case response for extreme conditions for the DER varying from zero to full output. Other levels such as the midday maximum and midday minimum for DER determine more probable bounds for the network/feeder response. The midday levels occur when DER is expected to generate a full output. The snapshot analysis procedure involves [52]:

- Solve the power flow with DER output set to zero.
- Lock all regulator and capacitor switches at their present state.
- Solve the power flow with DER generating the maximum power.

4.5.1.2 Time-Series Analysis

The time-series analysis is conducted for demand/generation coincident scenarios based on time of day. Basically, a time-of-day dependent load can be selected for the maximum load and a day for the minimum load. Each of these days should be evaluated by considering highly variable and non-variable DER outputs. The load and DER resemble scenarios that can be selected from the snapshot analysis results. The main objectives of the time series analysis are [52]:

- Determine feeder response from actual load and DER data.
- Compare time-series response to that indicated with the snapshot analysis.
- Determine DER influence on control elements.

An unbalanced time-series power flow simulation should be run considering changing the demand and generation profiles within a time (i.e., hourly, or less) and across various seasons. Therefore, time-series simulation can be carried out to consider the time variations of the profiles based on time-series data (i.e., load profiles, DER profiles, meteorological variables, etc.) with adequate granularity (e.g., at least 30-min resolution).

4.5.2 *Dynamic Power Flow*

Dynamic power flow (i.e., stability analysis) is essential in simulating the microgrids to ensure the stability of the system under several scenarios, including the variation of the loads and DERs over time. Dynamic power flow deals with power–frequency characteristics of DERs and loads to ensure the synchronisation and balancing in real-time between the demand-side and supply-side in the microgrid. Dynamic power flow involves the frequency response calculation with the power flow simulation to evaluate the stability of the microgrid associated with disturbances [53]. Therefore, dynamic models of the microgrid, including DERs, loads, controllers, switches, etc. are needed to secure accurate and realistic results.

4.5.2.1 Small-Signal Stability Analysis

Oscillations will be initiated simultaneously following the characteristics change of the electrical generation. In the explicit category of oscillation, the small signal instability occurs when the damping of one or more power oscillation modes of a power system is negative. The system is possible to be unstable in certain modes of operation. This situation is rarely seen in an intact microgrid, but it can occur when the system is heavily loaded by high power transmission or line interruptions due to disturbances and fault [54]. Persistent system oscillation with positive but small attenuation can also affect the stability of the power system, as they can also cause power line cascade outages due to incorrect trips in distance protection. They also lead to an undesired load on the turbine generator shaft during the power generation.

4.5.2.2 Transient Stability Analysis

Transient instability can occur in the microgrids between an individual synchronous machine or group of synchronous machines, as well as the rest components of the microgrids or between the microgrid and the main grid [55]. The typical reason for the occurrence is that power flow between an individual machine and the rest components of the network, or between two separate network areas that cannot be maintained due to a failure in one of the connecting lines or feeders.

4.5.2.3 Frequency Stability Analysis

The total active power generated, and the active power of the load consumption in the microgrid should always be balanced. If this criterion is failed to be met, the electrical frequency of the system will begin to change. In the case of overloaded scenarios, the synchronous machine is deprived of rotational energy, slowing down the electrical frequency. When the amount of electricity generated is larger than the load, the excess energy accelerates the machine and raises the electrical frequency [56].

4.5.2.4 Voltage Stability Analysis

Voltage instability usually occurs when the generators in microgrids are unable to supply (or consume) sufficient reactive power to maintain the system voltage within specified limits. This problem is often caused by one or more failures that are resolved by disconnecting a line or generator in the system. It is not uncommon for voltage instability to be a slow process affected by the operation of tap changers in the microgrid and the limits of the voltage regulators of the generator unit.

4.6 State Estimation

State estimation aims to estimate the states of the components in the network by minimising the weighted least squares between the actual measured network values and the variable values (e.g., node voltage phasor (i.e., voltage magnitude and phase angle), transformer turn ratios (i.e., turn ratio magnitude and phase shift angle), complex power flow (i.e., active, and reactive powers), active control status, etc.). Therefore, the accuracy of the estimated network variables depends on the accuracy of the measured values, the accuracy of the used model and the formulation of the problem. Several algorithms can be used to carry out the estimation, such as Gauss–Newton algorithm. One of the challenges of state estimation is the lack of data availability due to the limited measuring devices in the current networks. The required minimum number of measurement devices to minimise the estimation error state estimation is still an open research area. Distribution system state estimation (DSSE) has been proposed to handle the unbalanced systems recently applied in the "Increasing Visibility of Distribution Networks" project [57].

Summary

Model-based approach is the most used approach in Australia and worldwide for designing, planning, and operating distribution networks and microgrids based on the national and international reviewed projects. However, a data-driven approach can be a promising approach in modelling the microgrids as it can save time and resources, but the lack of data visibility is one of the main limitations of this approach as the accuracy of the results strongly depend on the volume and resolution of the data. In addition, it has several challenges in conducting "what-if" scenarios. Therefore, further research is needed in developing and validating the data-driven approach. Here, the selection of the models and the required data can be done based on the application and use case(s) as well as a trade-off between the accuracy, complexity, simulation burden and data availability.

Typically, the nodal power injections and branch power flow in a network are considered constant real and reactive power over the time interval (typically 5 or 30 min). Here, the power flow and state estimation should also consider two main constraints (i.e., voltage and thermal constraints) and the existing active energy management strategies (i.e., volt/var, volt/watt, etc.).

In this context, most of the reviewed national and international projects are using balanced steady-state power flow simulation. While the reaming projects did not clearly mention the type of power flow simulation used. However, unbalanced steady-state time-series simulations (e.g., power flow or state estimation) are essential to reflect the time-dependent aspects of demand/generation and controllable devices. This enables a more accurate assessment of voltages and power flows regarding time variation to support realistic dynamic "boundary conditions" in microgrid. This steady-state solution can be verified from an electromagnetic transient (EMT) simulation setup, either using dynamic power flow tools and/or hardware emulation systems.

4.7 Economic Analysis Factors

Economic analysis is typically a projection into the future, indicating action plans and their economic implications. It is not always a depiction of the experimental conditions or a cost–benefit analysis of the experiment. Rather, it is a research-based analysis that represents the actual implementation of microgrid projects outside of the demonstration framework. To guarantee adequate physical observations are obtained during the experimental demonstration phase, scoping the economic analysis—choosing what is to be included and what time frame it will be evaluated in—is critical [58].

4.7.1 Net Present Value

Net present value (NPV) is calculated as the investment costs in a project at the current time, including capital, annual operation and maintenance and any expected future costs/revenues in the cashflow during the life span of in a component in the microgrid or the whole microgrid [59]. The NPV can be expressed as:

$$NPV = \sum_{n=1}^{N} \frac{\textbf{\textit{Future cost in year n}}}{(1+d)^n} + \textbf{\textit{Initial cost}}, \tag{4.20}$$

where d stands for the discount rate, and N represents the period of analysis.

4.7.2 Annual Worth

Annual worth (AW) is an economic assessment of a project throughout its life cycle to support decision makers in selecting the optimum investment plan at the lowest cost. The AW is the total of yearly switching costs during the project's entire life cycle [60]. The AW can be expressed as:

$$AW = NPV \cdot \left(\frac{d \cdot (1+d)^N}{(1+d)^N - 1} \right). \tag{4.21}$$

4.7.3 Discounted Payback Period

The payback period is the number of years or time units required for a technology to recoup its initial and future costs when accounting for the time value of money. In other words, it is the amount of time necessary to accrue savings and revenues sufficient to offset the initial cost and any further future costs.

There are numerous mathematical formulas for calculating the deferred payback time. However, each formula is considered based on specific assumptions and should be utilised carefully based on the application and use case(s). It is recommended to monitor the cumulative cash flows in order to pinpoint the positive cash flow phase to accurately calculate the discounted payback period.

4.7.4 Levelised Cost of Energy

The levelised cost of energy (LCOE) measures the cost per unit of an energy produced across the microgrid component(s) over the assumed financial life span. It is utilised for investment planning and a fair comparison of different technologies with different lifetimes, project sizes, different capital costs, risks, returns and capacities [61]. The LCOE can be calculated as [62]:

$$LCOE = \frac{\textbf{Initial cost} + \sum_{n=1}^{N} \frac{\textbf{Cost in year } n}{(1+d)^n}}{\sum_{n=1}^{N} \frac{\textbf{Energy production in year } n}{(1+d)^n}}. \tag{4.22}$$

Summary

The economic analysis is one of the keystones in supporting the technical decision, where a high availability microgrid can be designed and operated at minimum cost. It is important to note that the discounted payback period should not be used in isolation from the NPV or AW as a decision criterion. Rather, it should be utilised as a supplemental measure to aid in decision making, as it may mislead the judgement. In economic evaluation, it is usually reported that the alternative with the shortest payback period does not necessarily produce the greatest economic value. This can also reflect on the community engagement in order to encourage them to invest in extending a microgrid project. Most of the microgrid projects are used net present value to estimate the investment cost of the design, while the annual total cycle cost is used in the operation test and the real-time energy management to reduce the operation costs of the microgrids. Here, the tariff structure and the FiT schemes should be included in the economic analysis to estimate the savings, the right value of the payback period as well as the support the decision of the offers and bids with the energy market.

References

1. Guerrero JM, Kandari R (2022) Microgrids: modeling, control, and applications, 1st edn. Elsevier
2. Australian Power Quality and Reliability Centre (2021) Microgrids: concepts and technologies. Wollomgong, NSW
3. Blackhall L (2020) On the calculation and use of dynamic operating envelopes. Canberra, ACT
4. Sub-Committee on MicroGrids Whitepaper, IEEE Smart Grid Technical Activities Committee (2022) Microgrids: utility challenges and opportunities
5. EPRI (2006) Distributed renewable energy generation impacts on microgrid operation and reliability. Palo Alto, California

6. Balderrama J, Gammie J, Rogers J (2021) Expanding the role of dispatchable DERs in US capacity markets
7. Office of Energy Efficiency and Renewable Energy—US Department of Energy (2022) Biopower. https://www.energy.gov/eere/bioenergy/biopower#:~:text=Biopowertechnologiesconvertrenewablebiomassfuelsintoheat,bacterialdecay%2Candconversiontogas%2Fliquidfuel. Burning. Accessed 15 Feb 2022
8. Rahman S, Pipattanasomporn M (2010) Modeling and simulation of a distributed generation-integrated intelligent microgrid. Arlington, VA
9. Delfino F, Procopio R, Rossi M et al (2018) Microgrid design and operation: toward smart energy in cities. Artech House
10. Hussain A, Sher HA, Murtaza AF, Al-Haddad K (2019) Improved restricted control set model predictive control (iRCS-MPC) based maximum power point tracking of photovoltaic module. IEEE Access 7:149422–149432
11. IEC (2006) IEC 61400-2: design requirements for small wind turbines
12. EPRI (2020) EPRI's energy storage roadmap, vision for 2025. Palo Alto, California
13. Hatziargyriou N, Asano H, Iravani R, Marnay C (2007) Microgrids. IEEE Power Energy Mag 5:78–94. https://doi.org/10.1109/MPAE.2007.376583
14. Economics and Industry Standing Committee (2019) Implications of a distribution energy future. Western Australia
15. Rajasekaran R, Rani PU (2021) Bidirectional DC-DC converter for microgrid in energy management system. Int J Electron 108:322–343. https://doi.org/10.1080/00207217.2020.1793418
16. Odo P (2020) A comparative study of single-phase non-isolated bidirectional dc-dc converters suitability for energy storage application in a dc microgrid. In: 2020 IEEE 11th international symposium on power electronics for distributed generation systems (PEDG). IEEE, pp 391–396
17. Guo Y, Meng J, Wang Y, Wang C (2019) A virtual DC machine control strategy for dual active bridge DC-DC converter. In: 2019 IEEE innovative smart grid technologies—Asia (ISGT Asia). IEEE, pp 2384–2388
18. Sim J, Lee J, Choi H, Jung J-H (2019) High power density bidirectional three-port DC-DC converter for battery applications in DC microgrids. In: 2019 10th international conference on power electronics and ECCE Asia (ICPE 2019—ECCE Asia). IEEE, pp 843–848
19. Erickson RW, Maksimović D (2020) Fundamentals of power electronics. Springer International Publishing, Cham
20. Singh B, Singh BN, Chandra A et al (2003) A review of single-phase improved power quality ac~dc converters. IEEE Trans Ind Electron 50:962–981. https://doi.org/10.1109/TIE.2003.817609
21. AEMO (2021) Application of advanced grid-scale inverters in the NEM
22. Hawaiian Electric Company (2020) Selling power to the utility. https://www.hawaiianelectric.com/documents/clean_energy_hawaii/selling_power_to_the_utility/competitive_bidding/20190822_final_stage_2_rfp_book_3.pdf. Accessed 3 Mar 2022
23. Ibrahim IA, Hossain MJ (2021) Low voltage distribution networks modeling and unbalanced (optimal) power flow: a comprehensive review. IEEE Access 9:143026–143084. https://doi.org/10.1109/ACCESS.2021.3120803
24. IEEE (2009) IEEE application guide for IEEE Std 1547(TM), IEEE standard for interconnecting distributed resources with electric power systems. IEEE Std 1547.2-2008 1–217
25. Standards Australia (2021) Grid connection of energy systems via inverters, Part 2: Inverter requirements. AS/NZS 4777.22020 Amd 12021
26. Ibrahim IA, Hossain MJ (2022) A benchmark model for low voltage distribution networks with PV systems and smart inverter control techniques. Renew Sustain Energy Rev 112571. https://doi.org/10.1016/j.rser.2022.112571
27. Braslavsky J, Graham P, Havas L et al (2021) N_2 opportunity assessment: low voltage network visibility and optimising DER hosting capacity: final report
28. Procopiou AT, Ochoa L (2019) Advanced planning of PV-rich distribution networks—deliverable 2: innovative analytical techniques. Victoria

29. Tian Y, Zhang F, Yuan Z et al (2020) Assessment power generation potential of small hydropower plants using GIS software. Energy Rep 6:1393–1404. https://doi.org/10.1016/j.egyr.2020.05.023
30. Tapia A, Reina DG, del Nozal AR, Millán P (2020) Application of genetic algorithms for designing micro-hydro power plants in rural isolated areas—a case study in San Miguelito, Honduras, pp 169–200
31. Dolatabadi SH, Ghorbanian M, Siano P, Hatziargyriou ND (2021) An enhanced IEEE 33 bus benchmark test system for distribution system studies. IEEE Trans Power Syst 36:2565–2572. https://doi.org/10.1109/TPWRS.2020.3038030
32. Mosa MA, Ali AA (2021) Energy management system of low voltage dc microgrid using mixed-integer nonlinear programing and a global optimization technique. Electr Power Syst Res 192:106971. https://doi.org/10.1016/j.epsr.2020.106971
33. Rowen WI (1983) Simplified mathematical representations of heavy-duty gas turbines
34. Hannett LN, Khan AH (1993) Combustion turbine dynamic model validation from tests. IEEE Trans Power Syst 8:152–158
35. Ibrahim IA, Hossain MJ, Duck BC, Fell CJ (2020) An adaptive wind-driven optimization algorithm for extracting the parameters of a single-diode PV cell model. IEEE Trans Sustain Energy 11:1054–1066. https://doi.org/10.1109/TSTE.2019.2917513
36. Ibrahim IA, Hossain MJ, Duck BC, Nadarajah M (2020) An improved wind driven optimization algorithm for parameters identification of a triple-diode photovoltaic cell model. Energy Convers Manag 213:112872. https://doi.org/10.1016/j.enconman.2020.112872
37. RWE npower renewables (2008) Wind turbine power calculations
38. Petru T, Thiringer T (2002) Modeling of wind turbines for power system studies. IEEE Trans Power Syst 17:1132–1139
39. Lei Y, Mullane A, Lightbody G, Yacamini R (2006) Modeling of the wind turbine with a doubly fed induction generator for grid integration studies. IEEE Trans energy Convers 21:257–264
40. Luo F, Dong ZY, Meng K et al (2016) Short-term operational planning framework for virtual power plants with high renewable penetrations. IET Renew Power Gener 10:623–633. https://doi.org/10.1049/iet-rpg.2015.0358
41. Salameh ZM, Casacca MA, Lynch WA (1992) A mathematical model for lead-acid batteries. IEEE Trans Energy Convers 7:93–98
42. Manwell JF, McGowan JG (1994) Extension of the kinetic battery model for wind/hybrid power systems. In: Proceedings of EWEC, pp 284–289
43. Rocha C, Peppanen J, Radatz P et al (2020) Storage element
44. Rocha C, Peppanen J, Radatz P et al (2020) storage controller element
45. Corrêa JM, Farret FA, Canha LN, Simoes MG (2004) An electrochemical-based fuel-cell model suitable for electrical engineering automation approach. IEEE Trans Ind Electron 51:1103–1112
46. Georgakis D, Papathanassiou S, Manias S (2005) Modeling and control of a small scale grid-connected PEM fuel cell system. In: 2005 IEEE 36th power electronics specialists conference. IEEE, pp 1614–1620
47. Luo R, Shafiee M (2021) The application of metaheuristics in optimal parameter identification of solid oxide fuel cell. Energy Rep 7:2563–2573
48. Khalilpour KR, Vassallo A (2016) A generic framework for distributed multi-generation and multi-storage energy systems. Energy 114:798–813. https://doi.org/10.1016/j.energy.2016.08.029
49. Zhou W, Mohammed N, Bahrani B (2022) Comprehensive modeling, analysis, and comparison of state-space and admittance models of PLL-based grid-following inverters considering different outer control modes. IEEE Access 10:30109–30146. https://doi.org/10.1109/ACCESS.2022.3157334
50. Sun J (2011) Impedance-based stability criterion for grid-connected inverters. IEEE Trans Power Electron 26:3075–3078. https://doi.org/10.1109/TPEL.2011.2136439
51. Suntio T, Messo T, Puukko J (2017) Power electronic converters: dynamics and control in conventional and renewable energy applications. Wiley

52. M. Rylander, Smith J (2012) Stochastic analysis to determine feeder hosting capacity for distributed solar PV. Knoxville, TN
53. Wang R, Zhang H, Li C, Ma C (2015) Dynamic power flow calculation method of power system with wind power. In: 2015 5th international conference on electric utility deregulation and restructuring and power technologies (DRPT). IEEE, pp 1987–1991
54. Kundur P, Paserba J, Ajjarapu V et al (2004) Definition and classification of power system stability. IEEE Trans Power Syst 19:1387–1401. https://doi.org/10.1109/TPWRS.2004.825981
55. Johansson N, Nee H-P (2011) Aspects on dynamic power flow controllers and related devices for increased flexibility in electric power systems
56. Anderson PM, Paul M, Fouad AA, Abdel-AA, Vittal V, McCalley JD. Power system control and stability
57. Terry J, Shafiei M, Krause O, Morosini G-M (2019) Solar enablement initiative project progress report. Queensland
58. Wakefield M (2012) Guidebook for cost/benefit analysis of smart grid demonstration projects, pp 1–118
59. IEEE Standards Association (2019) IEEE Std 2030.9-2019—IEEE recommended practices for the planning and design of the microgrid
60. Rao SS (2009) Engineering optimization: theory and practice, 4th edn
61. EIA (2013) Levelized cost of new generation resources in the annual energy outlook 2013. US Energy Inf Adm 1–5
62. Australian Energy Council (2022) Solar report

Chapter 5
Techno-economic Optimisation in Microgrids

5.1 General

Microgrids should be designed optimally to achieve the main goal(s) for use case(s). In which the components can be sized and located optimally based on one or a set of objective functions subject to one or a set of constraints. The objective functions and constraints may vary from one use case to another. Therefore, defining the use case is the keystone to identifying the relevant objective function(s) and constraint(s). In general, the microgrid planning and operation can be formulated and evaluated at the design stage. Accordingly, the optimisation to of microgrid's design can be split into three main stages: (i) planning; (ii) temporary islanded; and (iii) operation. The sequence of the design stages is important. In addition, if any of the objective function and/or constraint fails to converge, the planning scenario should be changed to make it converge. Here, operation optimisation aims to validate the solution proposed in both planning and temporary islanded stages under different conditions and scenarios either using historical data or so many random scenarios (e.g., time-series Monte Carlo simulation).

5.2 Techno-economic Optimisation

A techno-economic optimisation aims to find the optimal solution that provides a high availability at a minimum cost [1]. It typically uses mathematical models to estimate optimal objectives to meet the technical requirements by minimising cost and maximising benefits, such as investment cost, operating cost, and revenue based on technical and financial inputs (i.e., parameters, constraints) [2]. By linking process inputs to financial metrics, it can help business and industry better understand the factors that influence the profitability of their technology development projects.

I. A. Ibrahim et al., *Microgrids Design and Operation*, Power Systems,
https://doi.org/10.1007/978-981-96-4914-3_5

5.2.1 Optimisation for Microgrid Planning

In the case of a microgrid design, planners determine the location, size, and type of energy technology (both renewable and fossil fuels), ESSs and EVSE based on balancing the supply-side and demand-side [3]. The mathematical models of techno-economic optimisation can be generally divided into three main aspects, one is for planning purpose aiming to define the set of DERs to be installed [4], and the second is for ensuring that the proposed sizes and locations of DERs can operate in the islanded mode for a specific period of time and the other is for operation purpose aiming to determine the optimal schedule for power supply and storage for a particular topology configuration. Considering the technical limits, microgrids planning should comprehensively involve the DERs operating condition and system status, it should be implemented based on the application scenarios, objectives, system performance requirements, and cost/benefit analysis. From the mathematical perspective, these objective functions are expressed as one direction search (i.e., "min" or "max") equations of parameters and variables [5]. In generally, the techno-economic optimisation category is defined based on various purposes, it can be redefined based on other criteria. The other key entity in the optimisation is constraints, which represent technical (i.e., network and DER constraints), reliability, environmental, social, and legislative limits imposed for the solution of the optimisation problem. Examples of limitations in the microgrid planning problem are the available area (i.e., the roof of the building, the available area) within the area considered for the installation of solar and wind turbines. The last but most essential entity in the optimisation is variables, they are quantities that need to be determined and strategies that correspond to the decisions to be taken via solving the optimisation problem. For planning problems, examples of variables involve the size of DERs, location to install DERs, their numbers, type of DERs available in the microgrid, and the type and physical dimensions of the distributed system in the microgrid.

5.2.2 Power Generation Mix Selection and Sizing

The first consideration for planning is that the DERs, EVSE and energy storage equipment need to be sized corresponding to the highest demand (i.e., peak load) and economic criteria. Herein, not only minimising the initial investment can affect the design options, but other important factors may reflect in the decision, such as selecting the fuel type for the distributed microturbine if existed in the design option. In general, this step should be considered a strategic problem for the system. At the planning stage, three key goals need to be achieved, namely, cost-effectiveness, high reliability, and low impact on the environment.

5.2.3 Generation Sitting for Temporary Islanded

Usually, the planning problem aims to find the best location, size, number, type, and combination of available DERs, ESSs and EVSE technologies that can operate in grid-connected and islanded modes, taking into account estimated meteorological parameters (e.g., solar and wind power) under assumption or forecast, the energy demand, and several constraints [6]. Therefore, combing the operation problem optimisation problem for grid-connected and islanded modes together needs to include the short-term energy demand, frequent renewable output variation and energy prices. However, operation in this stage does not require strict accuracy as covering each temporal step data would bring computation task, typical average historical energy profile is normally used [7]. Thus, DERs, ESSs and EVSE must be sized and allocated optimally to maintain quality limits of balancing the demand-side and supply-side during the islanded operation mode. Planners must consider both critical/sensitive loads and non-critical/non-sensitive loads during this process (i.e., grid-connected, and islanded) to either supply the critical/sensitive loads or maximise the number of served loads based on the economic parameters and the islanded operation time. Consequently, feeders must serve customers areas, while being addressed to potential areas and operation modes. This stage can also be combined with the above strategic level problem. As with the sizing problem, the initial investment at this stage is directly related to the final design. This planning stage also requires high efficiency and reliability as in the previous plan, and a low power loss is also requested.

Following the reliability issue, the power networks/feeders have been suffering from increasing risks of outages due to the frequent occurrence of natural disasters and extreme weather events (e.g., bush fire, flood, and hurricane). Also, the rapidly growing energy demand can increase the risk of power outages [8]. As the DERs are a type of supply unit that can be deployed close to loads and invested in and planned by network enterprise. These network-owned DER units convert a microgrid from a passive to an active system, allowing the entire system to gain benefits by optimising its operation. Hence, the microgrid has the potential to improve power supply resilience. For instance, when a power outage occurs, corresponding protection devices would isolate the faults in the microgrid, but some out-of-service areas with no faults might face the power outage as well [9]. There is potential to utilise DERs to support and protect the critical/sensitive loads as far as possible after restoration. Therefore, the optimisation for this kind of load restoration is conducted to analyse the condition after the fault, but it can be considered as a kind of resilience capacity improvement priorly to increase the flexibility when facing the challenge in the future. Accordingly, if the temporary islanded design was not able to meet the design requirements and constraints for the specified time period, then the proposed planning scenario should be updated and evaluated again.

5.2.4 Optimisation for Microgrid Operation

There is a popular trend to co-optimise the planning and operation processes, by this way, it enables considering the power balance in the short-term (i.e., per hour or even less) for a long-term horizon. This design has a superiority that is to cover all the possible scenarios in the microgrid operation, especially in the islanded operation mode, while the main drawback is the increased complexity of computation. Further, the parameters in the techno-economic optimisation are, for instance, the energy efficiency of production generators and/or charging/discharging efficiency, or the availability of loads, DERs, ESSs and EVSE, capital cost or unit capacity cost for individual DERs, ESSs and EVSE technology. Besides, energy demand and DERs availability estimates should be available before running an optimisation problem. Also, energy prices should be set or forecasted as parameters, if considering the energy market trading behaviour.

After determining the optimal sizing scheme, the next step of microgrid operation would be complex as more participants are being involved. From the optimisation perspective, more relevant variables and constraints are introduced with the increasing number of participants. For example, ESSs are usually ignored, just like the physical dimensions of a grid, for the purpose of building simple decision-making problems based on planning problems, and sometimes it would be modelled as a fixed Investment cost or being integrated as extra cost into LCOE [10]. However, in the short-term operation problem, more explicit models are supposed to be built considering pursuing self-economic benefit via tariff arbitrage. Herein, the data are also required to be strictly consistent with the short-term temporal step (e.g., hourly changes are considered when there are trade-offs in energy strategy making).

In addition to the cost performance, the main objective of microgrid operation is to schedule the microgrid production to optimise specified key performance indicators, particularly, reliability. As microgrid would operate highly rely on the intermitted DERs (i.e., non-dispatchable DERs), which makes the reliability issue become more considerable on the short-term horizon. Although the presence of a connection to the upper grid always guarantees energy balance, normally the cost would increase with the reliability level enhancement. Given the relationship between reliability and economics, determining the ideal reliability level at which the operation procedure for enhancing system dependability is most cost-efficient in lowering customer damage costs. This kind of reliability involvement operation optimisation can be performed with different approaches, depending on the components of the microgrid, its topology, market rules, and optimisation goals. The common reliability assessment criterion is power loss, since it is directly relevant to the power balance, also it can reveal the reflection of technical variables to current microgrid operation. It is worth to point that the mentioned reliability criteria are certain indicators from the optimisation view, from other aspects, such as statistical analysis, other methods, such as Monte Carlo simulation can be applied to facing the reliability issue caused by uncertainties, further possibly considering probabilistic optimisation. Here, if one of the operation objective functions and/or considered constraints is failed, then the proposed planning scenario should be updated and passed the evaluation of the temporary islanded check to be assessed again using the operation requirements and constraints.

5.3 Social Aspects Incorporated into Optimisation Modelling

As previously mentioned, techno-economic optimisation for planning, temporary islanding and operating the microgrids are modelled through technical and economic aspects. The difference when using a socio-technical approach is incorporating social parameters without dismissing the importance of the techno-economic constraints [11]. Hence, design objectives beyond technical, cost and performance parameters must be included for the sustainable development of the microgrids. It is highly dependent on social interaction, which needs to be emphasised during the design process [12].

There are multiple challenges in modelling social dynamics and how they affect the energy demand over the lifetime of a microgrid. First, there is a need to understand how to incorporate the interaction between communities and energy systems into microgrid design. Since the characteristics of communities are distinctive, understanding what indicators should be included when assessing microgrid systems are essential. Therefore, social parameters will differ between communities. Priorities within each community will vary, and thus, measuring different forms of progress should be considered. Once the objectives of the energy system are determined, the associated social parameter(s) can be used in the design procedure and the assessment of the system. Each community will have a unique dynamic and set of challenges. Assessing those challenges and mapping them to energy access can support the decision-making progress [12].

One approach is to create identity factors from community members' relation to each other in the microgrid and the supply patterns from the DERs. Their identity factors could determine possible flexibility in each community member's behaviour. When modelling the planning, temporary islanding, and operating processes of the microgrid, the flexibility of each community member's consumption pattern is essential to realise as well. Depending on the member, smart-meter adaption may increase flexibility and generate information for the modelling stage of microgrid development [13]. This approach can help select which parameters are more important for different community members and how they can come into play with the planning, temporary islanded and operation. Hereby, incorporating social aspects in optimisation model leads to several benefits consist of:

- Energy independency.
- Energy security benefits.
- Energy efficiency.
- Localised air quality improvements.

- Greenhouse gas (GHG) emission reduction.
- Creating jobs: a reliable DERs, ESSs and EVSE industry will enable the creation of high skilled, high value jobs, thus decreasing the unemployment statistics.

5.4 Objective Functions

The formulation of the techno-economic optimisation stands for developing an optimisation problem aiming to optimise the size, location, and combination of available DERs, ESSs and EVSE to balance the supply and demand, considering technical, reliability, economic, and environmental requirements. A general mathematical representation of the techno-economic optimisation problem can be expressed as [14]:

$$\underset{x}{\boldsymbol{Min/Max}} \ \underbrace{f(\boldsymbol{x})}_{\textbf{Objective}}, \tag{5.1}$$

Subject to

$$\underbrace{h_i(\boldsymbol{x}) \le 0, i = 1, \ldots m,}_{\text{Inequality constraints}} \tag{5.1a}$$

$$\underbrace{g_j(\boldsymbol{x}) = 0, j = 1, \ldots n,}_{\text{Equality constraints}} \tag{5.1b}$$

$$x_{\min} \le \boldsymbol{x} \le x_{\max}, \tag{5.1c}$$

where $f(\boldsymbol{x})$: $\mathbb{R}^n \to \mathbb{R}$ denotes the objective; it is usually a function to minimise or maximise a goal over the variable vector $\boldsymbol{x} = [x_1, \ldots, x_i, \ldots, x_n]^T$. Herein, variables might include the size and the number of different types of DERs (e.g., solar PV, wind turbine, micro turbine, etc.), ESSs, EVSE, and so on. The terms $h_i(\boldsymbol{x})$ and $g_j(\boldsymbol{x})$ correspond to inequality and equality constraints. The microgrid's technical, economic, social, environmental, and physical aspects determine the objective and constraints. Several of the common objectives and constraints are described in the following subsections.

5.4.1 Technical Objectives

Technical objective functions aim to find the optimal size, settings, and location of the microgrid components based on the nature of the energy sources and the loads [15]. Here, the most used objective functions to address the technical aspects are listed and defined in Table 5.1.

Table 5.1 Commonly considered technical objectives

Objective function	Definition
Minimising total power losses	It aims to minimise the network loss, which is the difference between electrical power generated from embedded generators or the main grid, and electrical energy consumed [16]
Minimising line loss index	It aims to minimise the ratio of power loss to the power consumed by the loads [17]
Minimising total voltage deviation	It aims to minimise the summation of voltage deviations between buses in the microgrid and their reference voltages based on the standards [18]
Minimising reactive power demand	It aims to minimise the reactive power introduced to the network by developing energy storage control algorithms [19]
Maximising charging/discharging efficiency	It aims to maximise storage units' charging/discharging efficiency considering power and nominal efficiency related to ESSs [20]
Minimising active power curtailment	It aims to minimise the curtailment of active power produced by non-dispatchable DERs (e.g., PV systems and wind turbines) caused by unbalances between demand and generation, mainly during light-load periods [21]

5.4.2 Reliability Objectives

A microgrid's reliability represents its capability to satisfy the local electrical energy demand at all the time periods. An analysis of microgrid reliability indices is crucial when designing the microgrid, especially in the islanded operation mode [14]. The main aim of all reliability objectives is to ensure the balancing between the demand and supply at a specific time or at all time periods. The descriptions of the reliability objective functions are provided in Table 5.2.

5.4.3 Economic Objectives

Most of the economic objectives aim to minimise the microgrid costs, including the capital expenditure, cost of installation, operation, maintenance, replacement, and fuel costs of the backup generator. Maximising the revenue can also be formulated as an objective function [14]. The most used economic objectives are provided in Table 5.3.

Table 5.2 Commonly considered reliability objectives

Objective function	Definition
Minimising system average interruption duration index (SAIDI)	It aims to minimise the number of interrupted customers for each interruption event [22]
Minimising system average interruption frequency index (SAIFI)	It aims to minimise the number of interrupted customers for each interruption event associated with the restoration time for each interruption event [22]
Minimising momentary average interruption event frequency index (MAIFIe)	It aims to minimise the number of momentary interruptions per customer [22]
Minimising loss of load supply probability (LLSP)	It aims to minimise the probability of the energy resources failures to produce enough power to meet the demand [23]
Minimising loss of load probability (LLP)	It aims to minimise the probability of not meeting the load demands efficiently [24]
Minimising deficiency of power supply probability (DPSP)	It aims to minimise the deficit power amount in the microgrid [25]
Minimising expected energy not supplied (EENS)	It aims to minimise the expected energy that the microgrid will not supply to the loads for a given period [26]

Table 5.3 Commonly considered economic objectives

Objective function	Definition
Minimising total microgrid cost	It aims to minimise the sum of all the costs incurred in the process of installation, maintenance, operation, replacing, and components fuel costs within the microgrid [27]
Minimising capital cost	It aims to minimise the cost associated with acquiring fixed assets and making long-term investments such as buying equipment and upgrading technology [16]
Maximising net present value (NPV)	It aims to maximise the cash flow profits of the microgrid by adding up all incomes and expenses, considering interest and discount rates [28]
Minimising the allocation cost	It aims to minimise the different DERs configuration costs depending on the location and natural environment of a microgrid [17]
Minimising annual total life-cycle cost (ATLCC)	It aims to minimise the total annual costs of the maintenance, operation, replacement and fuel costs of the microgrid components [24]

(continued)

Table 5.3 (continued)

Objective function	Definition
Minimising maintenance cost	It aims to minimise the cost of asset management, equipment maintenance, routine inspections, reporting, regular system integration and reviews, and preventive maintenance [29]
Minimising total operational cost	It aims to minimise the operational costs of the microgrid once it has been installed [30]
Minimising generation cost	It aims to minimise the cost of the total power produced by all the generating units [16]
Minimising fuel cost	It aims to minimise the overall fuel costs of the power produced by all the non-renewable generators in the microgrid [31]
Minimising levelised cost of energy	It aims to minimise the ratio of the total microgrid cost to the total amount of energy generated [25]
Minimising shortage cost	It aims to minimise the costs associated with economically compensating consumers if the microgrid cannot satisfy their demand during a particular time period. This can incentivise consumers to postpone demand until another time [32]
Minimising start-up and shut-down costs	It aims to minimise the costs associated with switching the generators on and off [21]
Minimising hourly capital cost of energy storage systems	It aims to minimise the capital cost of an energy storage system, which is determined by the storable energy and the peak power that required to be delivered [33]
Minimising storage cost	It aims to minimise the charging and discharging cost of energy storage system, which is calculated based on charge and discharge powers, cost of ESS for periodic duty, and operation state of controllable units [34]
Minimising hourly operating and maintenance cost of energy storage system	It aims to minimise operating and maintenance costs of the ESSs, which is determined by the hourly discharged energy and rated power [33]
Minimising fuel cells cost	It aims to minimise the costs related to the operation and maintenance of fuel cell and costs of degradation effects due to fluctuating profiles of fuel cell's power [35]

(continued)

Table 5.3 (continued)

Objective function	Definition
Minimising cost of ultra-capacitor	It aims to minimise the costs of ultracapacitors that maintained at an intermediate state of charge in order to balance a deficit between short-term and long-term schedules [36]
Minimising hydrogen storage costs	It aims to minimise the functional cost of hydrogen regulation service, including penalty for final deviation, life cycle costs, degradation profile costs, operating and maintenance costs, and start-up costs [36]
Minimising frustration cost	It aims to minimise a frustration cost that occurs when an appliance is not turned on when it is needed the most [37]
Minimising dis-satisfaction cost	It aims to minimise the costs due to the deviating from nominal energy consumption results in the cost of dissatisfaction [38]
Minimising penalty cost	It aims to minimise the penalty for inaccurate PV energy forecasts submitted to the utility company. An energy supplier will compensate for the inconvenience caused by unrealised supply bids [39]
Minimising demand response cost	It aims to minimise the payments of demand response customers, including capacity offered and actual demand response loads [21]
Minimising load-shedding cost	It aims to minimise the penalties that will be imposed if the involuntary load is curtailed [21]
Minimising payment of controllable loads	It aims to minimise the hourly price signals to schedule the energy consumption of controlled loads, which aims to reduce the cost of the power purchased [40]
Maximising benefits at lower time-of-use prices	It aims to maximise the profit from the difference in energy prices between light-load and peak-load periods, which includes the effect of levelling load based on time-of-use rates. It is essential to schedule energy storage charge/discharge operations so that low-priced energy is stored during light-load periods, and then delivered during peak-load periods [33]
Maximising microgrid revenue	It aims to maximise the revenues derived from the purchase and sale of electricity to the main grid [34]
Minimising worst-case net transaction cost	It aims to minimise the microgrid's worst-case transaction costs, which are driven by its ability to trade surplus energy with the main grid and import energy when shortages occur [41]

(continued)

Table 5.3 (continued)

Objective function	Definition
Minimising cost of importing energy from other microgrids	It aims to minimise the overall cost of purchasing energy from the other microgrids [42]
Minimising purchase cost from the utility grid	It aims to minimise the overall cost of purchasing energy from the main grid [43]
Minimising grid cost	It aims to minimise the costs to penalise deviations from the long-term economic dispatch schedule for the purchase and sale of energy from the main grid [36]

5.4.4 Environmental Objectives

One of the key reasons for promoting microgrids is to reduce GHG emissions caused by electricity production [44]. Microgrids can include backup systems, such as diesel generators with greenhouse emissions to increase reliability. The environmental impacts can affect the decision making, funding and social acceptance. In designing a microgrid, reducing emissions is one of the most common objectives [14]. The environmental objectives are listed in Table 5.4.

Table 5.4 Commonly considered environmental objectives

Objective function	Definition
Minimising greenhouse emission cost	It aims to minimise the total cost of emissions generated by the microgrid over its lifetime [45]
Minimising levelised emission	It aims to minimise the amount of GHG emissions released from the microgrid to the total electricity generated in a given year [23]
Minimising carbon emissions	It aims to minimise the quantity of carbon emissions produced by the DERs in the microgrid [16]
Minimising carbon emissions with penalty cost	It aims to minimise the penalties imposed for CO_2 emissions by DERs in the microgrid [16]
Minimising carbon emissions related to imported power	It aims to minimise the quantity of CO_2 emissions produced by the DERs in the microgrid and emissions related to the imported power from the main grid, especially the grey power (i.e., not the green power) [46]

5.5 Constraints

A major consideration when formulating the microgrid problem is the constraints, which define the technical and reliability limits of the microgrid and the integrated DERs, considering the economic, environmental and social constraints [47]. Some of the objective functions can be considered as constraints and vice versa, which depends on the problem formulation, such as some of the reliability and economic constraints. Details of common constraints are provided in this subsection.

5.5.1 Network Constraints

Several network constraints should be considered in planning, temporary islanding and operating the microgrids. The parameters of network constraints should be set by the microgrid operator based on the regulations and standards. The network constraint can be listed in Table 5.5.

Table 5.5 Network constraints

Constraint	Definition
Overvoltage and undervoltage limit	This criterion is the primary concern of the distributed energy system, where the intrusion of too many renewable energy sources in the wrong place can lead to an unacceptable rise in bus voltage [48]
Voltage deviation	The rapid increase in output voltage adversely affects the voltage regulation devices in the distribution system, and consumers also complain about the voltage deviation
Voltage imbalance	The voltage imbalance per phase can be determined as a percentage deviation from the average at each node with penetration for renewable energy sources [48]
Thermal limits	It is an important limiting factor that limits the total renewable energy capacity to the given feed-in. High DER penetration at off-peak conditions can cause massive reverse power flow [48]
Protection	It is an important limiting factor that limits the total DER capacity to the given feed-in. High off-peak DER penetration can cause massive backflow. This constraint includes the total fault current, forward flow fault current, breaker reduction of reach, sympathetic tripping, and anti-islanding [48]
Power factor	The power factor is one of the power quality constraints. The power factor should be limited by maximum and minimum values as well as if it is lagging or leading based on the standards

(continued)

Table 5.5 (continued)

Constraint	Definition
Resonance	It is a power quality constraint and depends on the total number of capacitors integrated in the branch [48]
Distortion	It is a power quality constraint. The impact of DER distortion is dependent on the four variables, including phase angle of harmonics, configuration of capacitors, load level and size and location of DER [48]
Control	The control limits are analysed in the simulations. A time series analysis is carried out on each protective device for various cases with an increased duty cycle. The capacitors and regulators in the network, performing more than one operation at a time, reduce the life of the device [48]
Frequency stability	It stands for the variation of the frequency of a crystal oscillator in the microgrid because of some external conditions, such as temperature variation, voltage variation, output load variation, and frequency aging
Rotor angle stability	It is the ability to maintain synchronism after being subjected to a disturbance, which represents the torque balance of synchronous machines
Voltage stability	It is the ability to maintain steady acceptable voltage at all buses in the system under normal operating conditions and after being subjected to a disturbance
Renewable energy fraction	Describes the share of renewable energy in the microgrid compared with non-renewable energy sources [25]
Reverse power flow	The ability of the microgrids to export the energy to the main grid [30]
Energy generation profile	Meteorological parameters determine how much energy is generated from renewable sources. Microgrid design is affected by this constraint since it considers the maximum amount of energy that can be generated at a given time [25]
Load demand balance	There should be a balance between DERs generation and demand in the microgrid [49]

5.5.2 *Reliability Constraints*

Various reliability measures have been developed for quantifying the microgrid's ability to meet the load demand [14]. Some of the reliability indices can be considered as a constraint and an objective. Here, Table 5.6 lists the most common reliability constraints in design the microgrids.

5.5.3 *Distribution Energy Resources Constraints*

DER constraints have been considered in the optimisation to ensure that the DERs in planning, temporary islanded and operation conditions are within the limits during the recognition of the best location and size of candidate sources. Therefore, the explicit

Table 5.6 Commonly considered reliability constraints

Constraint	Definition
System average interruption duration index (SAIDI)	It measures the number of interrupted customers for each interruption event [22]
System average interruption frequency index (SAIFI)	It measures the number of interrupted customers for each interruption event associated with the restoration time for each interruption event [22]
Momentary average interruption event frequency index (MAIFIe)	It measures the number of momentary interruptions per customer [22]
Loss of load supply probability (LLSP)	It refers to the proportion of time for which the microgrid fails to power the loads [23]
Loss of load probability (LLP)	It measures how often a microgrid or energy system does not meet load demands efficiently [24]
Deficiency of power supply probability (DPSP)	It measures how likely it is that the microgrid will fail to meet the energy demand. The reason can be a system breakdown or a low level of power generation from renewable energy sources [25]
Expected energy not supplied (EENS)	It measures the expected energy that the microgrid will not supply for a given period [26]
Loss of load expected (LOLE)	It measures the average number of hours when load demand exceeds microgrid generation [50]
Equivalent loss factor (ELF)	It is the ratio of the load outage time to the total operating time [51]
Total energy loss (TEL)	It is an indication of the energy loss caused by the extra power generated in the microgrid [52]
Loss of load risk (LOLR)	It is a measure of the probability that a deficiency in DERs can cause the failure of the system to meet daily load requirements [53]
Loss of load hours (LOLH)	It represents the total annual hours that the microgrid cannot meet the load due to deficiency in the generation or a reduced energy supply [54]
System performance level (SPL)	It is a measure of how many days the microgrid is not capable of meeting the load demand [55]
Unmet load (UL)	It calculates the ratio of un-served load to the total microgrid load in a year [53]
Net dump energy (NDE)	It measures the dumped energy when the microgrid's generation exceeds its load, which may happen in islanded mode [56]
Loss of energy probability (LEP)	It refers to the ratio of the waste of energy in the microgrid to the overall energy demand in a year [57]
Percentage of health state probability (PHSP)	It measures the percentage of time that the microgrid has sufficient reserves to meet the defined reserve criteria [58]
Risk state probability (RSP)	It measures the percentage of time when the system cannot meet the load, thus exhibiting a risk [58]

(continued)

Table 5.6 (continued)

Constraint	Definition
Energy fluctuation rate (EFR)	It measures how well the microgrid matches generation and load demand. This can be analysed based on the closeness between generation and load curves [57]
Total blackout duration (TBD)	It represents microgrid outages time because of a power shortage, no matter how severe it might be [58]
Mean time between failures (MTBF)	It represents the ratio of overall time when the microgrid operation has been successful over a period of time to the number of system failures. The system is successful when its load demand is less than or equal to its available power [58]

DER constraints are listed in this subsection for the most common technologies in microgrids.

5.5.3.1 Small-Scale Hydropower Plants Constraints

The potential of small-scale hydropower plants and their contribution in providing the power to the microgrids, as a dispatchable DER, depend on several parameters and constraints. Therefore, several small-scale hydropower plants constraints should be considered in designing the microgrids as listed in Table 5.7.

5.5.3.2 Small Reciprocating Internal Combustion Engines Constraints

Small reciprocating internal combustion engines represent several types of fuel engines. These engines are considered as dispatchable DERs, which can help in stabilising the system and ensure the balancing between the demand and supply. To ensure the proper operation of these engines, several constraints should be considered in the design stage. These constraints are listed in Table 5.8.

5.5.3.3 Microturbine Constraints

Several microturbine constraints should be considered in the optimisation to ensure that microturbine potential and contribution to the microgrid are within the limits during the allocation and sizing selection recognition. The microturbine constraints are defined in Table 5.9.

Table 5.7 Commonly considered small-scale hydropower plants constraints

Constraint	Definition
Power output	The power output depends on the location, which varies based on the location, hydraulic head, generator type, turbine type, the control of the operation and water discharge curve
Expected lifetime	It is the measure of the average time small-scale hydropower plants are expected to be utilised, based on the year when it is produced, daily use, and other factors like maintenance
Number limit	The number of small-scale hydropower plants should not exceed the maximum number of small-scale hydropower plants. Such limitation is referred to as the number of small-scale hydropower plants limitation
Size/Capacity limit	This kind of constraint limits the size of a single small-scale hydropower plant. Following the size limitation, the capacity of planned small-scale hydropower plant should not exceed the maximum capacity
Power relevant limit	• The power factor should be kept within a proper range • The active power output of a small-scale hydropower plant should be kept within the range between the lower bound and the upper bound • The reactive power output of a small-scale hydropower plant should be kept within the range between lower bound and the upper bound
Ramping limit	Ramp-up and ramp-down rate should be limited within the range between the lower bound and the upper bound, respectively Ramp-up and ramp-down duration should not exceed the maximum time, respectively

Table 5.8 Commonly small reciprocating internal combustion engines constraints

Constraint	Definition
Power output	The power output depends on the generator type, fuel type, fuel curve, and the control of the operation
Expected lifetime	It is the measure of average time small reciprocating internal combustion engines are expected to be utilised, based on the year when it is produced, daily use, and other factors like maintenance
Number limit	The number small reciprocating internal combustion engines should not exceed the maximum number of small reciprocating internal combustion engines. Such limitation is referred to as the number of small-scale hydropower plants limitation
Size/Capacity limit	This kind of constraint limits the size of a single small reciprocating internal combustion engine. Following the size limitation, the capacity of planned small reciprocating internal combustion engine should not exceed the maximum capacity
Power relevant limit	• The power factor should be kept within a proper range • The active power output of small reciprocating internal combustion engines should be kept within the range between the lower bound and the upper bound • The reactive power output of small reciprocating internal combustion engines should be kept within the range between the lower bound and the upper bound
Ramping limit	• Ramp-up and ramp-down rate should be limited within the range between the lower bound and the upper bound, respectively • Ramp-up and ramp-down duration should not exceed the maximum time, respectively

Table 5.9 Commonly considered microturbine constraints

Constraint	Definition
Power output	The power output depends on the generator type, fuel type, fuel curve, and the control of the operation
Expected lifetime	It is the measure of average time microturbines are expected to be utilised, based on the year when it is produced, daily use, and other factors like maintenance
Number limit	The number of microturbines should not exceed the maximum number of microturbines. Such limitation is referred to as the number of microturbines limitation
Size/ Capacity limit	This kind of constraint limits the size of a single microturbine. Following the size limitation, the capacity of the planned micro turbine should not exceed the maximum capacity
Power relevant limit	• The power factor should be kept within a proper range • The active power output of microturbine should be kept within the range between the lower bound and the upper bound • The reactive power output of microturbine should be kept within the range between the lower bound and the upper bound
Ramping limit	• Ramp-up and ramp-down rate should be limited within the range between the lower bound and the upper bound, respectively • Ramp-up and ramp-down duration should not exceed the maximum time, respectively

5.5.3.4 Solar Photovoltaic System Constraints

The potential, location and size of the PV systems depend on several factors and constraints. Therefore, several constraints should be considered to ensure the realistic operation of the PV systems and their contribution to balancing the demand and supply sides. These constraints are listed in Table 5.10.

5.5.3.5 Wind Turbine Constraints

The selection of wind turbine location, size, and operation in the microgrid can be optimised based on several constraints to ensure the optimally selection. Therefore, the commonly considered wind turbine constraints are listed in Table 5.11.

5.5.3.6 Electrical Storage Battery Constraints

Electrical storage batteries are one of the most important microgrid components, especially in the islanded operation mode to ensure the balancing between the demand and supply sides. Therefore, selecting the proper capacity and location of electrical storage batteries affect the operation and the control of microgrid and its services. Therefore, several constraints should be considered to optimally size, allocate and operate the electrical storage batteries as listed in Table 5.12.

Table 5.10 Commonly considered PV system constraints

Constraint	Definition
Power output	The power output depends on the type of PV module, solar radiation\temperature curve, and the control of the operation
Expected lifetime	It is the measure of the average time PV panels are expected to be utilised, based on the year when it is produced, daily use, and other factors like maintenance
Number limit	The number of PV panels should not exceed the maximum number of PV panels. Such limitation is referred to as the number of PV panels limitation
Size/Capacity limit	The size of a single PV panel is restricted. Following the size limitation, the capacity of planned PV panels should not exceed the maximum capacity
Penetration level limit	Being relevant to number, size, and capacity, the maximum penetration of PV panels, the maximum PV penetration on each bus, and the maximum allowable penetration of wind turbines in the entire network are limited, due to the reliability consideration
Power relevant limit	• The power factor should be kept within a proper range • The active power output of PV system should be kept within the range between the lower bound and the upper bound • The reactive power output of PV system should be kept within the range between the lower bound and the upper bound
Ramping limit	• Ramp-up and ramp-down rate should be limited within the range between the lower bound and the upper bound, respectively • Ramp-up and ramp-down duration should not exceed the maximum time, respectively

Table 5.11 Commonly considered wind turbine constraints

Constraint	Definition
Power output	The power output depends on the type of wind turbine, length of the turbine, generator type, wind speed curve, and the control of the operation
Expected lifetime	It is the measure of the average time wind turbines are expected to be utilised, based on the year when it is produced, daily use, and other factors like maintenance
Number limit	The number of wind turbines should not exceed the maximum number of wind turbines. Such limitation is referred to as the number of wind turbines limitation
Size/Capacity limit	The size of a single wind turbine is restricted. Following the size limitation, the capacity of planned wind turbines should not exceed the maximum capacity
Penetration level limit	Being relevant to number, size, and capacity, the maximum penetration of wind turbines, the maximum wind turbine penetration on each bus, and the maximum allowable penetration of wind turbines in the entire network are limited, due to the reliability consideration

(continued)

Table 5.11 (continued)

Constraint	Definition
Power relevant limit	• The power factor should be kept within a proper range • The active power output of wind turbine should be kept within the range between lower bound and the upper bound • The reactive power output of wind turbine should be kept within the range between lower bound and the upper bound
Ramping limit	• Ramp up and ramp down rate should be limited within the range between lower bound and the upper bound, respectively • Ramp up and ramp down duration should not exceed the maximum time, respectively

Table 5.12 Commonly considered electrical storage battery constraints

Constraint	Definition
Expected lifetime	It is the measure of the average time batteries are expected to be utilised, based on the year when it is produced, daily charging and discharging, and other factors like maintenance
Number limit	The number of batteries should not exceed the maximum number. Such limitation is referred to as the number of batteries limitation
Size/Capacity	This kind of constraint limits the size of a single battery. Following the size limitation, the capacity of planned battery should not exceed the maximum capacity
Charging and discharging power limit	• The charging power should neither exceed the upper bound nor be less than the lower bound • The discharging power should neither exceed the upper bound nor be less than the lower bound
SOC limit	The SOC should be kept within the range between lower bound and the upper bound
Initial and end state	The initial and end state for battery to conduct charging and discharging should be limited
Depth of discharge (DOD) limit	It is the percentage of the total capacity of the battery, and it represents how much energy is cycled into and out of the battery during a charging and discharging cycle
Power relevant limit	• The power factor should be kept within a proper range • The active power output of electrical battery should be kept within the range between lower bound and the upper bound • The reactive power output of electrical battery should be kept within the range between lower bound and the upper bound

(continued)

Table 5.12 (continued)

Constraint	Definition
Ramping limit	• Ramp up and ramp down rate should be limited within the range between lower bound and the upper bound, respectively • Ramp up and ramp down duration should not exceed the maximum time, respectively

5.5.3.7 Hydrogen Fuel Cell Constraints

Hydrogen fuel cell is a promising technology in microgrids. Several factors and constraints should be followed during the planning, temporally islanded and operation stages to ensure the optimal sizing, allocating and operating of HFC in the microgrid. The commonly used hydrogen fuel cell constraints are defined in Table 5.13.

Table 5.13 Commonly considered hydrogen fuel cell constraints

Constraint	Definition
Power output	The power output depends on the fuel cell type, fuel type, fuel curve, and the control of the operation
Expect lifetime	It is the measure of average time fuel cells are expected to be utilised, based on the year when it is produced, daily use, and other factors like maintenance
Number limit	The number of fuel cells should not exceed the maximum number of hydrogen fuel cell. Such limitation is referred to as the number of hydrogen fuel cells limitation
Size/ Capacity	This kind of constraint limits the size of a single fuel cell. Following the size limitation, the capacity of planned fuel cell should not exceed the maximum capacity
Thermal limit	It limits the thermal energy of fuel cell keep within a proper range between lower bound and upper bound
Electricity limit	It limits the electricity produced by fuel cells to keep within a proper range between lower bound and upper bound
Flowing fuel	It is to control the fuel consumption by fuel cells, it can also be formulated with the help of pressure limits to control the volume, and flowing speed of fuel
Power relevant limit	• The power factor should be kept within a proper range • The active power output of the fuel cell should be kept within the range between the lower bound and the upper bound • The reactive power output of the fuel cell should be kept within the range between the lower bound and the upper bound
Ramping limit	• Ramp-up and ramp-down rate should be limited within the range between the lower bound and the upper bound, respectively • Ramp-up and ramp-down duration should not exceed the maximum time, respectively

5.5.3.8 Electric Vehicles Supply Equipment Constraints

EVSE consists of the EVs and charging/discharging stations. Modelling and operating the EVSE are sensitive processes in Microgrids, especially during the temporary islanded operation mode. The charging and discharging of EVs can affect the power balance and power quality in microgrid. Therefore, several constraints should be considered in optimising the EVSE as listed in Table 5.14.

Table 5.14 Commonly considered electric vehicles supply equipment constraints

Constraint	Definition
Capacity	The energy limit for EVs, which means the planned energy for EVs should not exceed the maximum capacity
Charging and discharging power limit	The charging power of EVs should neither exceed the upper bound nor be less than the lower bound The discharging power of EVs should neither exceed the upper bound nor be less than the lower bound
State of charge (SOC) limit	The SOC of EVs should be kept within the range between lower bound and the upper bound
Initial and end state	The initial and end state for EVs to conduct charging and discharging should be limited
Charging and discharging duration	The charging and discharging time duration should not exceed the EVs owner's expected time horizon
Depth of discharge (DOD) limit	It is the percentage of the total capacity of the battery in the EVs, and it represents how much energy is cycled into and out of the battery in the eVs during a charging and discharging cycle
Power relevant limit	• The power factor should be kept within a proper range • The active power output of the EVSE should be kept within the range between the lower bound and the upper bound • The reactive power output of EVSE should be kept within the range between the lower bound and the upper bound
Ramping limit	• Ramp-up and ramp-down rate should be limited within the range between the lower bound and the upper bound, respectively • Ramp-up and ramp-down duration should not exceed the maximum time, respectively

5.5.4 Economic Constraints

The economic factors are one of the most important aspects in the techno-economic optimisation. Accordingly, the economic aspects could not be involved in the formulation of the objective function; therefore, they should be considered as constraints. Here, the economic factors can be considered as objective functions and as constraints as mentioned in this subsection [59]. The commonly considered economic constraints are listed in Table 5.15.

Table 5.15 Commonly considered economic constraints

Constraint	Definition
Capital cost	It represents one-time expenses incurred on the investment for equipment such as DER, feeders, and transformers. In other words, capital cost denotes the total cost required to make the project executable
Maintenance cost	It refers to the amount of money that needs to be spent to provide the proper maintenance of microgrid components. These costs could be yearly
Replacement cost	It refers to the amount of money currently spent on replacing essential equipment like a PV panel to keep the power balance
Depreciation cost	It refers to the depreciation of the microgrid components
Marginal cost	It is the change in total power cost that comes from producing one additional unit. It is normally used in operation problem
Budget constraints	Budget constraint is the total amount of equipment that can be affordable within a current budget. Budget constraint illustrates the range of choices available within that budget. It can be limited as inequality constraint with a bound
Tariff constraints	It is the effort of consumers or reward of prosumer to pay or sell the tariff for the electricity. It can be limited as an inequality constraint with a bound
Fuel cost constraints	It refers to the specific type of fuel cost during the planning and operation. It can be limited as an inequality constraint with bound
Ancillary services cost constraints	It refers to the economic cost of purchasing or economic benefits of providing ancillary services. It can be limited as inequality constraint with a bound, the cost can be further limited based on explicit services such as the payment for congestion relief, voltage control, reactive power compensation, and frequency regulation
Payback period	It refers to the time that is required to return the investment cost

5.5.5 *Social Constraints*

The social impact on microgrid projects strongly affect the decision-making in the design. Therefore, incorporating social parameters is essential to ensure the engagement and participation of the community members in designing, temporary islanding, operating, and upgrading the microgrid projects. Some of the economic, reliability and environmental constraints can affect the incorporation of social aspects in microgrid projects. The lifecycle cost of microgrids with the consideration of utility bill savings, regulation penalties, cost of energy production and use, and social variables such as social cost of carbon and public health cost can be considered as social constraints [60]. Accordingly, the commonly used social constraints are listed in Table 5.16.

Further on, parameters considering community engagement, flexibility and satisfaction have been used as microgrid modelling constraints in previous research. These parameters have been addressed through general system modelling or analysed separately.

5.6 Microgrid Design Optimisation Model

An optimisation framework is being developed to select the most reliable and the most competitive investment alternative(s) for various electricity generation technologies. The model of the optimisation process is based on a large number of iterations to

Table 5.16 Commonly considered social constraints

Constraint	Definition
Regulation penalties	They represent the penalties that the microgrid operator should be paid to the prosumers for the issues related to power quality (e.g., voltage limits, power factor, etc.), curtailment (e.g., active power, charging/discharging), maximum demand, and reliability
Social cost of carbon	The climate benefits and reduction of social damages are based on reducing CO_2 emissions from saved non-renewable electricity purchased from the grid or on-site production [61]
Loss of load probability (LLP)	The impact of power outages or unstable electricity delivery can cause numerous losses that affect business interruption, damage to machinery, loss of productivity and even loss of life. Can be expressed a quantified through economic measures [11]
Job creation	Jobs created through microgrid deployment, energy planning installation and operation process, and increased resilience of community through jobs/MW [60]
CO_2 emissions	The individual emission factor is estimated through a life cycle assessment for the power generation in the microgrid. This can be used to compare with conventional components and fossil fuel grids [60]

find the life-cycle cost (LCC) of several energy generation technologies to provide a balanced solution that secures a high level of reliability at lowest LCC.

5.6.1 Optimisation Model

The optimisation framework is designed based on LCC as the critical variable for the optimisation outcome. Intrinsically, life cycle costing is an economic appraisal technique that is widely used to compare and measure the economic performance of different investment alternatives. It is based on the concept of time value of money, which is a critical threshold for comparing different investment alternatives with relatively long future time horizon.

5.6.1.1 Objective Function

The total LCC includes all costs that will be incurred throughout the service life of a system. It considers initial acquisition cost and installation, operating cost including running cost, minor and major repair, and replacement cost, in addition to end-of-life cost. The LCC for a system can be estimate for the whole life cycle of the product or system, or it can be limited by a predefined period of analysis. Several factors affect the period of analysis, among others, economic life, functional life, and technological life.

Life cycle savings and benefits are also essential components of LCC analysis. The objective function for the optimisation framework is determined based on the Net life cycle costs, benefits, and saving. Two measures are commonly used to quantify LCC, the NPV, and the AW. While the first represents all costs and benefits in the present money value, the latter is based on the concept of annual equivalence of all costs and benefits. Technically, the optimisation framework finds solutions based on the following formula:

$$\textbf{Maximise } AW = G^{MG} - C^{T}, \tag{5.2}$$

where G^{MG} stands for the annual equivalent value of total gain, and C^{T} represents the annual equivalent value of total cost.

Designing an optimisation framework for various renewable energy generation technologies should, inevitably, take into account two majors' factors: the first factor is technology reliability to meet energy demand, while the second factor is the economic viability of the technology. Bearing these two thresholds in mind, the model architecture has been designed to identify the best solution that meets the highest technical and economic performance.

5.6.1.2 Constraints

The objective function is designed to be solved subject to several network and DER constraints to ensure the technical deployment of the DERs in the microgrid. Accordingly, the technical part of the techno-economic optimisation framework is modelled based on the following constraints.

5.6.1.3 Basic Constraints

The basis constraint aims to ensure that the selected capacity and the apparent power flow for DER k that connected to bus b in phase a at time t are within the pre-defined limits.

Maximum DER capacity constraint

This constraint aims to ensure the optimised capacity of a DER k, at bus b is within the DER k predefined maximum capacity in kVA, which can be represented as:

$$\mathbf{0} \leq s_{kb} \leq S_k, \tag{5.3}$$

where s_{kb} is a variable that represents the capacity of DER k at bus b, and S_k is a parameter stands for the maximum capacity of DER k.

DER power flows constraint

The power flow is subject to an apparent power flow limit of each DER, which is limited by the nominal apparent power rating that can be mathematically formulated as:

$$\sum_{a \in \Xi_b} p_{abkt}^2 + q_{abkt}^2 \leq s_{kb}^2, \tag{5.4}$$

where a is an index represents the phase in the set of phases of bus b (Ξ_b), p_{abkt} is a variable stands for the active power injected by DER k, in phase a at bus b in time t, and q_{abkt} is a variable stands for the reactive power injected by DER k, in phase a at bus b in time t.

5.6.1.4 Voltage Constraints

The voltage constraint aims to limit the voltage magnitude in phase a at bus b within time t ($|v_{abt}|$). Accordingly, the $|v_{abt}|$ should be limited between a minimum and maximum voltage range as:

$$V^{min} \leq |v_{abt}| \leq V^{max} \tag{5.5}$$

where $V^{\min}$ and $V^{\max}$ represent the minimum and maximum limits of the allowed voltages (in Australia $V^{\min} = 0.94\,p.u.$ and $V^{\max} = 1.1\,p.u.$ based on AS 60038).

5.6.1.5 Thermal Constraints

The current constraint aims to limit the current magnitude flows in phase a at bus b through the conductor c within time t ($|i_{abct}|$). Accordingly, the $|i_{abct}|$ should be set as:

$$|\boldsymbol{i}_{abct}| \le \boldsymbol{I}_{c}^{max} \tag{5.6}$$

where $I_c^{\max}$ is the maximum current limit of conductor c (ampacity).

5.6.1.6 Three-Phase and Single-Phase DER Constraints

This constraint aims to ensure that if the DER k is a three-phase or single phase. Accordingly, if the DER k is a three phase, then the power injected to the three-phase bus b must be balanced. Otherwise, the power injected can follow the constraints relevant to the bus b.

To check the DER k is a single-phase or three-phase, the following formulas should be followed:

$$-S_k \gamma_{akb} \le p_{aktb} \le S_k \gamma_{akb}, \tag{5.7}$$

$$-S_k \gamma_{akb}, \le q_{aktb} \le S_k \gamma_{akb}, \tag{5.8}$$

$$\sum_{a \Xi_b} \gamma_{\{akb\}} = \gamma_{\{kb\}}^{\{AUX\}} + 3\left(1 - \gamma_{\{kb\}}^{\{AUX\}}\right), \tag{5.9}$$

where $\gamma_{\{akb\}}$ is a binary variable of node use for DER k connected to bus b at phase a, and $\gamma_{\{kb\}}^{\{AUX\}}$ is a binary variable of connection of the DER k at bus b (if $\gamma_{\{kb\}}^{\{AUX\}} = 1$ means single-phase, while if $\gamma_{\{kb\}}^{\{AUX\}} = 0$ it means three-phase).

If DER k is three-phase, then the power injected must be balanced based on the following constraints:

$$-S_k \gamma_{\{kb\}}^{\{AUX\}} \le p_{\{1ktb\}} - p_{\{2ktb\}} \le S_k \gamma_{\{kb\}}^{\{AUX\}}, \tag{5.10}$$

$$-S_k \gamma_{\{kb\}}^{\{AUX\}} \le p_{\{2ktb\}} - p_{\{3ktb\}} \le S_k \gamma_{\{kb\}}^{\{AUX\}}, \tag{5.11}$$

$$-S_k \gamma_{\{kb\}}^{\{AUX\}} \le p_{\{3ktb\}} - p_{\{1ktb\}} \le S_k \gamma_{\{kb\}}^{\{AUX\}}, \tag{5.12}$$

where $p_{\{1ktb\}}$ is the active power injected by DER k to bus b in phase 1 at time t, $p_{\{2ktb\}}$ stands for the active power injected by DER k to bus b in phase 2 at time t, and $p_{\{3ktb\}}$ is the active power injected by DER k to bus b in phase 3 at time t.

5.6.1.7 Life Cycle Cost and Benefits Model

As outlined earlier, the LCC and benefits model is designed to predict the life cycle costs and benefits along a predefined period of analysis. For ease of use, the model allows users to define a desired period of analysis based on their discretion.

The LCC and benefits model is applied through an optimisation framework. Therefore, the basic constraints of the DERs and networks have been imposed on the model to be incorporated in the optimisation framework. Therefore, as an initial step and based on the electricity maximum load, the optimisation framework is designed to estimate the total hourly electrical energy generation that is required to meet the peak demand. To do so, the optimisation framework performs check to decide whether to go through a standalone technology, or through a combination of technologies. For the optimiser to reach a final decision, it is designed to follow the above-mentioned objective function through estimating the LCC and benefits associated with each alternative and base the decision on the highest LCC value as an objective function as discussed on Sect. 5.6.1.1.

On hourly basis, the optimisation quantifies: (1) the featured energy by the costumer in the first year to find the amount of excess energy that can be sold to generate income, (2) the amount of annual energy saving, and (3) the amount of black (non-green) energy in case of shortage.

Energy demand growth rate, in addition to degradation rates of technologies highly impacts the future amounts of featured energy. For example, the growth rate increases the energy demand, which reduces the amount of excess energy exported to the grid, as well as it reduces the annual savings. The same argument applies for the degradation rate. Accordingly, the optimisation framework is designed to take both the growth and the degradation into account in making future projects for each component of the featured energy.

To quantify the annual featured energy by the costumer considering degradation and growth rates, the optimisation framework follows the following steps:

- First year active and reactive energy generation by technologies

$$E_{bk}^{A} = \sum_{a \in \Xi_b} \sum_{t \in 8760} p_{abkt}, \tag{5.13}$$

$$E_{bk}^{R} = \sum_{a \in \Xi_b} \sum_{t \in 8760} q_{abkt}, \tag{5.14}$$

where E_{bk}^{A} stands for the active energy generated by DER k at bus b in the first year in kWh, and E_{bk}^{R} is the reactive energy generated/consumed by DER k at bus b in the first year in kVARh.

- Annual active and reactive energy generation by DER $\boldsymbol{k}$ with annual degradation effect

$$\boldsymbol{E_{bk\psi}^{A}} = \overline{\boldsymbol{d_{k\psi}}} \cdot \boldsymbol{E_{bk}^{A}}, \tag{5.15}$$

$$\boldsymbol{E_{bk\psi}^{R}} = \overline{\boldsymbol{d_{k\psi}}} \cdot \boldsymbol{E_{bk}^{R}}, \tag{5.16}$$

where $E_{bk\psi}^{A}$ stands for the active energy generated by DER k at bus b in year ψ in kWh, $E_{bk\psi}^{R}$ represents the reactive energy generated/consumed by DER k at bus b in year ψ in kVARh, and $\overline{d_{k\psi}}$ is the annual compounded degradation factor for DER k in year ψ in p.u.

- First year actine and reactive energy consumption by the costumer

$$\boldsymbol{E_{b}^{ACT}} = \sum_{a \in \Xi_b} \sum_{t \in 8760} \boldsymbol{P_{abt}^{CT}}, \tag{5.17}$$

$$\boldsymbol{E_{b}^{RCT}} = \sum_{a \in \Xi_b} \sum_{t \in 8760} \boldsymbol{Q_{abt}^{CT}}, \tag{5.18}$$

where E_{b}^{ACT} stands for the active energy consumed by the customer at bus b in the first year, E_{b}^{RCT} is the reactive energy consumed by the customer at bus b in the first year, P_{ab}^{CT} stands for the active power demand by the customer connected to bus b through phase a at time t, and Q_{ab}^{CT} is the reactive power demand by the customer connected to bus b through phase a at time t.

- Annual energy consumption by the customer with annual growth effect

$$\boldsymbol{E_{b\psi}^{ACT}} = (\boldsymbol{1} + \boldsymbol{g})^{\boldsymbol{\psi}} \cdot \boldsymbol{E_{b}^{ACT}}, \tag{5.19}$$

$$\boldsymbol{E_{b\psi}^{RCT}} = (\boldsymbol{1} + \boldsymbol{g})^{\boldsymbol{\psi}} \cdot \boldsymbol{E_{b}^{RCT}}, \tag{5.20}$$

where $E_{b\psi}^{ACT}$ stands for the active energy consumed by the customer connected at bus b in year ψ in kWh, $E_{b\psi}^{RCT}$ represents the reactive energy consumed by the customer connected at bus b in year ψ in kVARh, and g is the annual energy demand growth rate in p.u.

- Featured energy by the costumer in time ($\boldsymbol{t}$) in the first year

$$\boldsymbol{P_{bt}^{NET}} = \sum_{a \in \Xi_b} \sum_{k \in K} \boldsymbol{p_{abkt}} - \sum_{a \in \Xi_b} \boldsymbol{P_{abt}^{CT}}, \tag{5.21}$$

where P_{bt}^{NET} stands for the active power difference between the active power generated by all the DERs the connected to bus b at time t and the active power demand by the customer connected to bus b at time t in kW.

- And finally, affected by degradation and growth rates annual featured energy by the costumer

$$\boldsymbol{E_{b\psi}^{excess}} = \sum_{k \in K} (\mathbf{1} - \boldsymbol{g})^{\boldsymbol{\psi}} \cdot \overline{\boldsymbol{d_{k\psi}}} \cdot \boldsymbol{E_{b}^{excess}}, \tag{5.22}$$

$$\boldsymbol{E_{b\psi}^{saving}} = \sum_{k \in K} (\mathbf{1} - \boldsymbol{g})^{\boldsymbol{\psi}} \cdot \overline{\boldsymbol{d_{k\psi}}} \cdot \boldsymbol{E_{b}^{saving}}, \tag{5.23}$$

$$\boldsymbol{E_{b\psi}^{deficit}} = \sum_{k \in K} (\mathbf{1} + \boldsymbol{g})^{\boldsymbol{\psi}} \cdot \left(\overline{\boldsymbol{d_{k\psi}}}\right)^{\boldsymbol{deficit}} \cdot \boldsymbol{E_{b}^{deficit}}, \tag{5.24}$$

where $E_{b\psi}^{excess}$ stands for the energy excess (exported to the grid) at bus b in year ψ in kWh, E_{b}^{excess} is the active energy excess at bus b in the first year in kWh, $E_{b\psi}^{saving}$ represents the energy savings, which is the met demand by the DERs at bus b in year ψ in kWh, E_{b}^{saving} is the energy savings, which is the met demand by the DERs at bus b in the first year in kWh, $E_{b\psi}^{deficit}$ is the energy shortage (imported from the grid) for the customer at bus b in the year ψ in kWh, $E_{b}^{deficit}$ represents the energy shortage (imported from the grid) for the customer at bus b in the first year in kWh, and $\left(\overline{d_{k\psi}}\right)^{deficit}$ stands for the annual compounded deficit degradation factor (it has an opposite effect compared with $\overline{d_{k\psi}}$) for DER k in year ψ in p.u.

The aforementioned equations represent the technical part of the optimisation model. Applying the above formulas yields the following results:

- Peak load in kW
- System capacity in kW
- Annual and life cycle electricity generation considering degradation rates in kWh
- Annual and life cycle excess electricity generation considering degradation rates in kWh
- Annual and life cycle electricity saving considering growth rate in kWh
- Annual and life cycle black energy considering growth rate in kWh

The above list, technically, contains the required technical input to conduct LCC analysis to evaluate the economic viability of the various technology alternatives that meet the technical requirements.

Estimating the LCC for each alternative is highly affected by two major factors: the discount rate and the inflation rate. While the first, in its simple form, refers to money productivity, the later refers to the decrease in the purchase power of money. The optimiser exhibits high flexibility by allowing the user to, independently, input the desired real discount rate, the expected general price inflation, fuel escalation rates, in addition to electricity prices escalation rate. The rationale behind this flexibility is the fact that fuel prices escalate at different levels than that of electricity and general price inflation rates. Hence, for accurate and realistic future cost and budget projections, the varying escalation rates cannot be overlooked.

The renewable energy technologies are wide and varied, therefore, it is a daunting task to predict future inflation rates for each technology in terms of major repair and replacement cost. For this reason, dealing with inflation is limited to fuel and electricity price. In other words, the general inflation rate is used for future cost projects other than the fuel and electricity cost, which are projected using separate escalation rates. it worth mentioning that the user has the option to eliminate the effect of price inflation by using zero value as an input for inflation field.

The optimisation framework calculates the inflation-adjusted discount rate, or sometimes referred to as nominal discount rate as follow:

- General inflation/deflation adjusted discount rate (j):

$$j = (i + f) + (i \cdot f), \tag{5.25}$$

  where i stands for the real interest/discount rate, and f represents the general inflation/deflation rate in p.u.
- Fuel escalation adjusted discount rate (Z): (The fuel escalation rate applies for all fuel types based on each technology.)

$$Z = (i + z) + (i \cdot z), \tag{5.26}$$

  where z is the annual fuel escalation rate in p.u.
- Electricity escalation adjusted discount rate (X):

$$X = (i + x) + (i \cdot x), \tag{5.27}$$

  where x represents the annual electricity price escalation rate in p.u.

It is noteworthy that, both fuel and electricity future costs are discounted based on its respective adjusted discount rate. This, mathematically, means that both fuel and electricity escalations rates have the function to estimate future cost projections; they have no effect in the discounted LCC.

As discussed earlier, both the NPV and the net AW can be utilised as measures of the LCC and benefits. It is decided to adopt the AW as a measure for the LCC. The rationale behind this decision is the limitation of the NPV in comparing technologies with varied service life. Furthermore, the AW is easier in terms of interpreting the results, and more meaningful for comparing different alternatives with annual cost and benefits.

The optimisation framework estimates the annual value of the cost components based on the following:

- Annual equivalent value of the initial costs

The optimisation framework is designed to model information related for various types of renewable energy generation technologies namely: rooftop and large-scale PV systems, residential and large-scale electrical storage batteries, hydropower, wind turbine, in addition to diesel generator option although it is not classified as a renewable energy technology. The rationale behind including diesel generator is to resolve any reliability issues that might be encountered as a result of employing a renewable energy technology.

The optimisation framework is designed to estimate the annual equivalent value (AW) of the initial cost of setting up renewable energy technologies, in addition to other initial cost implications associated with land acquisition cost and required transformer upgrade to accommodate renewable energy sources. As an optimisation framework, land rent or buy decision is reached by the optimisation framework based on user input for land rent costs. Land acquisition cost, as a buy option, is estimated by the optimisation framework based on predefined criteria that are incorporated in the framework for each technology.

Similarly, the optimisation framework is designed to check whether a transformer upgrade is required or not. A threshold of 75 kVA as a system capacity is predefined in the framework. Accordingly, the optimisation framework is capable to estimate the required additional initial cost to upgrade the current transformers if the system capacity is required to be more than 75 kVA. The formula followed by the optimisation framework to find the initial cost with transformer upgrade is as follow:

$$C_k^I = \left(\sum_{b \in B} C_k^U \cdot s_{kb} + \sum_{w \in W} C_w^{TRU} \right) \cdot \left(\frac{j \cdot (1+j)^{\mu}}{(1+j)^{\mu} - 1} \right) + \text{Land acquisition cost}, \tag{5.28}$$

where C_k^I strands for the annual equivalent value of the initial costs for DER k in \$, C_k^U is the Initial investment unit costs of DER k in \$/kVA, s_{kb} represents the capacity of DER k at bus b in kVA, C_w^{TRU} is the Initial investment unit costs of transformer w in \$, and μ stands for the period of analysis in year.

While if the transformation upgrade is not required, the framework applies the following:

$$C_k^I = \sum_{b \in B} C_k^U \cdot s_{kb} \cdot \left(\frac{j \cdot (1+j)^{\mu}}{(1+j)^{\mu} - 1} \right) + \textbf{Land acquisition cost}. \tag{5.29}$$

Accordingly, the total annual equivalent value of the initial costs for all the DERs (C^I) can be obtained as follows:

$$\boldsymbol{C^I} = \sum_{\boldsymbol{k \in K}} \boldsymbol{C_k^I}. \tag{5.30}$$

- Annual equivalent value of the replacement costs

Considering that each renewable energy technology has its own replacement schedule, the optimisation framework user has to input replacement costs as future inflated/deflated costs. However, due to the fact that predicting the future replacement cost for each technology is a daunting task due to inflation/deflation effect. Therefore, to generalise the usability of the framework, future replacement costs are discounted using the real discount rate. The real discount rate is a user input, and hence, the user still has the option to adjust the real discount rate to accommodate certain amount of expected inflation rate.

The optimisation framework estimates the annual value of the replacement cost, including transformer replacement, based on the following formula:

$$\boldsymbol{C_k^R} = \left(\sum_{\boldsymbol{b \in B}} \sum_{\boldsymbol{\psi \in \Psi_k}} \overline{\boldsymbol{C_{k\psi}^R}} \cdot \boldsymbol{s_{kb}} \cdot (\boldsymbol{1} + \boldsymbol{i})^{-\boldsymbol{\psi}} + \sum_{\boldsymbol{w \in W}} \sum_{\boldsymbol{\psi \in \Psi_w}} \overline{\boldsymbol{C_{w\psi}^{TRR}}} \cdot (\boldsymbol{1} + \boldsymbol{i})^{-\boldsymbol{\psi}} \right) \cdot \left(\frac{\boldsymbol{j} \cdot (\boldsymbol{1} + \boldsymbol{j})^{\boldsymbol{\mu}}}{(\boldsymbol{1} + \boldsymbol{j})^{\boldsymbol{\mu}} - \boldsymbol{1}} \right) \tag{5.31}$$

where C_k^R strands for the annual equivalent value of the replacement costs for DER k in \$, $\overline{C_{k\psi}^R}$ is the future replacement costs of DER k for different replacement years ψ in \$/kVA, where $\psi \in \Psi_k$, $\boldsymbol{\Psi_k}$ is a set of DER k replacement years, $\overline{C_{w\psi}^{TRR}}$ represents the future replacement costs of transformer w for different replacement years ψ in \$, where $\psi \in \Psi_w$, and Ψ_w is a set of transformer w replacement years.

If transformer replacement does not apply, the following formula is used:

$$\boldsymbol{C_k^R} = \sum_{\boldsymbol{b \in B}} \sum_{\boldsymbol{\psi \in \Psi_k}} \overline{\boldsymbol{C_{k\psi}^R}} \cdot \boldsymbol{s_{kb}} \cdot (\boldsymbol{1} + \boldsymbol{i})^{-\boldsymbol{\psi}} \cdot \left(\frac{\boldsymbol{j} \cdot (\boldsymbol{1} + \boldsymbol{j})^{\boldsymbol{\mu}}}{(\boldsymbol{1} + \boldsymbol{j})^{\boldsymbol{\mu}} - \boldsymbol{1}} \right). \tag{5.32}$$

Accordingly, the total annual equivalent value of the replacement costs for all the DERs (C^R) can be found as follows:

$$\boldsymbol{C^R} = \sum_{\boldsymbol{k \in K}} \boldsymbol{C_k^R}. \tag{5.33}$$

- Annual equivalent value of the operation and maintenance (O&M) costs

Considering that the annual operating and maintenance costs for each technology represents the average annual O&M costs throughout the period of analysis, which includes minor and major repairs, the optimisation framework follows the following formula to estimate the O&M costs including transformer O&M costs:

$$C_k^{O\&M} = \sum_{\psi=1}^{\mu}\left(\sum_{b\in B} C_k^{O\&M} \cdot s_{kb} + \sum_{w\in W} C_w^{TRO\&M}\right) \cdot \left(\frac{(1+f)^{\psi}}{(1+j)^{\psi}}\right) \cdot \left(\frac{j\cdot(1+j)^{\mu}}{(1+j)^{\mu}-1}\right), \tag{5.34}$$

where $C_k^{O\&M}$ strands for the annual equivalent value of the O&M costs for DER k in \$, $C_k^{O\&M}$ is the O&M costs of DER k in \$/kVA, and $C_w^{TRO\&M}$ is the O&M costs of transformer w in \$.

If transformer O&M does not apply, the following formula is used:

$$C_k^{O\&M} = \sum_{b\in B}\sum_{\psi=1}^{\mu} C_k^{O\&M} \cdot s_{kb} \cdot \left(\frac{(1+f)^{\psi}}{(1+j)^{\psi}}\right) \cdot \left(\frac{j\cdot(1+j)^{\mu}}{(1+j)^{\mu}-1}\right) \tag{5.35}$$

It is noteworthy that, the optimisation framework estimates the annual operating and maintenance cost as inflated cost subjected to general price inflation rates that is defined by the user, and hence, the framework is design to apply inflation adjusted discount rate to find the annual worth of O&M costs. Accordingly, the total annual equivalent value of the O&M costs for all the DERs $\left(C^{O\&M}\right)$ can be calculated as follows:

$$C^{O\&M} = \sum_{k\in K} C_k^{O\&M}. \tag{5.36}$$

- Annual equivalent value of other general future costs (optional)

In order to increase the flexibility and reliability of the optimisation framework, the option to define miscellaneous future costs is offered by the framework. The use may wish to define future costs for a certain technology other than O&M, and replacement costs. The following formula is used if additional future cost is defined:

$$C_{\psi}^{FC} = \overline{C_{\psi}^{FC}} \cdot (1+i)^{-\psi} \cdot \left(\frac{j\cdot(1+j)^{\mu}}{(1+j)^{\mu}-1}\right), \tag{5.37}$$

where C_{ψ}^{FC} strands for the annual equivalent value of other general future costs in \$, and $\overline{C_{\psi}^{FC}}$ is the other general future costs in year in ψ in \$.

Accordingly, the total annual equivalent value of all other future costs during the period of analysis $\left(C^{FC}\right)$ can be found as follows:

$$C^{FC} = \sum_{\psi\in\underline{\Psi}} C_{\psi}^{FC}. \tag{5.38}$$

- Annual equivalent value of the fuel costs

Although the fuel costs can be considered under the annual O&M costs, the optimisation framework is designed to estimate the fuel cost separately due to the fact that the average annual escalation rate of fuel costs is expected to be significantly different than that of other annual O&M costs as well as not all the technologies have fuel costs. As mentioned earlier, the optimisation framework allows the user to define fuel escalation rate beside the general inflation rate, and hence, the framework uses the user-defined fuel escalation rate in estimating future fuel cost for relevant technology (C^F) as follows:

$$\boldsymbol{C^F} = \sum_{k\in K}\sum_{b\in B}\sum_{\psi=1}^{\mu}\left(\boldsymbol{E_{kb\psi}} \cdot \boldsymbol{C_k^F}\right) \cdot \left(\frac{(\mathbf{1}+z)^{\psi}}{(\mathbf{1}+\boldsymbol{Z})^{\psi}}\right) \cdot \left(\frac{\boldsymbol{j} \cdot (\mathbf{1}+\boldsymbol{j})^{\mu}}{(\mathbf{1}+\boldsymbol{j})^{\mu}-\mathbf{1}}\right), \tag{5.39}$$

where $E_{kb\psi}$ strands for the annual energy generated by DER k that requires fuel connected at bus b in the year ψ in kVAh, and C_k^F is the fuel costs for DER k in $/kVAh.

- Annual equivalent value of the salvage gain

The salvage value (future resell value) that is associated with end of life, and technology, or its parts replacements is classified as a gain associated with each technology. The optimisation framework allows the user to input the percentage of each technology salvage value from its initial investment/replacement cost, then, the framework estimates the corresponding salvage value at the end of life, or at a certain replacement year. Several scenarios exist for the salvage value, and accordingly, the optimisation framework follows several conditional statements to estimate the salvage value in the future. The framework follows the following general formula to estimate the salvage value in a certain year based on applicable conditional statement.

$$\boldsymbol{G_k^S} = \sum_{b\in B}\boldsymbol{\lambda_k} \cdot \boldsymbol{C_k^U} \cdot \boldsymbol{s_{kb}} \cdot (\mathbf{1}+\boldsymbol{i})^{-\tau_k} \cdot \left(\frac{\boldsymbol{j} \cdot (\mathbf{1}+\boldsymbol{j})^{\mu}}{(\mathbf{1}+\boldsymbol{j})^{\mu}-\mathbf{1}}\right) \tag{5.40}$$

where G_k^S strands for the annual equivalent value of the salvage gain for DER k in $, λ_k is the percentage of DER k salvage value from the initial investment/replacement cost in %, and τ_k stands for the service life of DER k in years.

Accordingly, the total annual equivalent value of the salvage gains from all the DERs $\left(G^S\right)$ can be found as follows:

$$\boldsymbol{G^S} = \sum_{k\in K}\boldsymbol{G_k^S}, \tag{5.41}$$

Finally, in estimating the annul equivalent value of all LCC components, the optimisation framework aggregates the results obtained from the above formulas to yield annual equivalent value of all the system costs and salvage values as follows:

$$C^T = C^I + C^R + C^{O\&M} + C^{FC} + C^F - G^S, \tag{5.42}$$

It is worth mentioning that the optimiser is designed to estimate the cost components in isolation from the benefits, therefore, the above equations do not include the revenues and cost savings. Although the optimiser has been designed to estimate the net annual cash flows, then, calculate the net annual worth, it has been decided to build the framework to deal with the benefits in isolation from the costs. This, however, secures higher reporting flexibility for users in terms of future cost and revenue budgeting.

Similar to the cost components, the optimiser follows the following steps to estimate the revenues, cost saving, and the annual electricity bill cost (in case of shortage).

- Annual equivalent value of the revenue generated by all technologies and energy bill savings

The revenue generated from renewable energy technologies is, simply, dependent of the amount of excess energy and the feed-in tariff (FiT). While the monetary value of the saving is determined by the amount of energy saved and the retailer price. The optimisation framework estimates the revenue and savings based on the following formula, which considers user input for electricity price escalation:

$$\begin{aligned} G^{RS} = &\sum_{b \in B} \sum_{\psi=1}^{\mu} F^{FiT} \cdot E_{b\psi}^{excess} \cdot \left(\frac{(1+x)^{\psi}}{(1+X)^{\psi}} \right) \cdot \left(\frac{j \cdot (1+j)^{\mu}}{(1+j)^{\mu} - 1} \right) \\ &+ \sum_{b \in B} \sum_{\psi=1}^{\mu} F_b^{CT} \cdot E_{b\psi}^{saving} \cdot \left(\frac{(1+x)^{\psi}}{(1+X)^{\psi}} \right) \cdot \left(\frac{j \cdot (1+j)^{\mu}}{(1+j)^{\mu} - 1} \right), \end{aligned} \tag{5.43}$$

where G^{RS} stands for the annual equivalent value of the revenue generated by all DERs and energy bill savings through the period of analysis μ in \$, F^{FiT} stands for the FiT rate in \$/kWh, and F_b^{CT} is the energy retailer rate at bus b in \$/kWh.

It is worth noting that in the above formula, the future escalated energy revenue and savings are discounted using an electricity inflation adjusted discount rate, and hence, mathematically, the annual equivalent value of revenues and saving is only affected by the amount of general price inflation as determined by the user of the framework. The electricity escalation rate is used to quantify future revenues and savings for financial budgeting and reporting only.

- Annual equivalent value of the energy bill cost

The optimisation framework is designed to estimate energy deficit in case that the renewable energy technology failed to meet the entire demand. Therefore, the bill cost is estimated based on the amount of energy deficit and retailer price, in addition to the daily supply charge by utility service providers. The following formula applies to find the annual equivalent value of the bill cost after considering electricity price inflation.

$$C^{EB} = \sum_{b\in B} \sum_{\psi=1}^{\mu} \left(\left[F_b^{CT} \cdot E_{b\psi}^{deficit} \cdot \left(\frac{(1+x)^{\psi}}{(1+X)^{\psi}} \right) \cdot \left(\frac{j \cdot (1+j)^{\mu}}{(1+j)^{\mu} - 1} \right) \right] + \left[(F_b^S \cdot 365) \cdot \left(\frac{(1+x)^{\psi}}{(1+X)^{\psi}} \right) \cdot \left(\frac{j \cdot (1+j)^{\mu}}{(1+j)^{\mu} - 1} \right) \right] \right), \tag{5.44}$$

where C^{EB} represents the annual equivalent value of the energy bill cost for all the customers in the network through the period of analysis μ in \$, and F_b^S stands for the daily supply charge for the customer at bus b in \$/day.

- Annual equivalent value of all revenues, savings, and energy bill costs

Finally, the bill cost is deducted to quantify the net revenues and savings as follow.

$$G^{MG} = G^{RS} - C^{EB},$$

Finally, the optimiser generates results based on the following objective function:

$$\textbf{Maximise } AW = G^{MG} - C^{T}. \tag{5.45}$$

5.6.2 *Optimisation Solver*

The unbalanced OPF model requires a solver to handle the optimisation in the unbalanced power equations, objective function and the set of constraints to find the optimal solution. Several commercial and open-access solvers can be use in this context. As the developed optimisation framework aims to be generalised, which can be used by any DNSP without any extra cost, an open-access optimiser is utilised, which is called interior point optimiser (Ipopt) [62].

Ipopt is using the interior point algorithm to solve the optimal power flow problems, which aims to guarantee convergence by using line search methods, trust region constraints, techniques to handle non-convexity via merit functions, and other modifications [62]. Ipopt employs an interior point line search filter method to locate a local solution in non-linear programming. For more details about interior point algorithms and various extensions, the reader can refer to [63–66]. It is considered more efficient solver in terms of performance and iteration counts compared with the other solvers that using the interior point algorithm (e.g., LOQO and Knitro). It is also widely used by several open access tools that solving the unbalanced non-linear OPF problems, such as Open-DSOPF and PowerModelsDistribution.

The mathematical representation of the interior point algorithm that used by Ipopt solver is explained in detail in [62, 67–70], which can be employed to solve general non-linear programming as follows:

$$\min \boldsymbol{f}(\boldsymbol{x}), \tag{5.46}$$

subject to

$$\boldsymbol{g}^{\boldsymbol{L}}(\boldsymbol{x}) \le \boldsymbol{g}(\boldsymbol{x}) \le \boldsymbol{g}^{\boldsymbol{U}}, \tag{5.46a}$$

$$\boldsymbol{x}^{\boldsymbol{L}} \le \boldsymbol{x} \le \boldsymbol{x}^{\boldsymbol{U}}, \tag{5.46b}$$

where f: $R^n \rightarrow R$ is the objective function, $x \in R^n$ represent the optimisation variables, $x^L \in (R \cup \{-\infty\})^n$ and $x^U \in (R \cup \{+\infty\})^n$ are the lower bounders and upper bounders of x, respectively, g: $R^n \rightarrow R$ stands for the general non-linear constraints, and $g^L \in (R \cup \{-\infty\})^m$ and $g^U \in (R \cup \{+\infty\})^m$ are the lower bounders and upper bounders for the constraint g, respectively.

Note that $f(x)$ and $g(x)$ can be linear or non-linear and convex or non-convex but should be twice continuously differentiable. In addition, the equality constraints of the form $g_i(x) = \overline{g}_i$ can be specified by setting $g_i^L = g_i^U = \overline{g}_i$ [51].

Summary

An optimisation framework has been developed based on a well-defined LCC objective function and subject to DERs and network constraints to select the most reliable and cost-effective investment alternative(s) for various dispatchable and non-dispatchable DERs, as well as for electrical storage batteries and EVSE. The objective function is formulated based on the time value of money, which is a crucial criterion for comparing investment alternatives with a reasonably long-time horizon in the future. A system's LCC might be estimated for the entire product or system's life cycle, or it can be confined to a predetermined time of analysis. Both the NPV and the AW can be used to quantify the costs and benefits over the life cycle. As a measure of the LCC, the AW is selected. This choice was made due to the inadequacy of the NPV when comparing technologies with different service lives. In addition, the AW makes it simpler to evaluate the data and provides a more realistic basis for evaluating various alternatives based on their annual costs and benefits. The optimisation framework is expressed as an imbalanced OPF problem, requiring a non-linear solution to solve the unbalanced OPF equations. An open-access, Ipopt, solver is employed because the built optimisation framework is intended to be applicable by any DNSP at no extra expense.

5.7 Microgrid Optimisation Framework

Microgrids should be optimally designed to achieve the main aims based on one or several use cases. In this context, the DERs, electrical storage batteries and EVSE can be optimally selected based on their potential and optimally allocated and sized for their capacity based on one or more objective functions and a set of constraints. The objective functions and the set of constraints may differ based on the use case(s).

In general, microgrid planning and operation can be established and assessed during the design phase to ensure the optimal selection and operation of the components during the period of analysis. This should be done based on the load/generation profiles, DER costs and service of life, and energy costs, and subject to technical constraints (e.g., network constraints, DER constraints, etc.).

Consequently, optimising the microgrid's design can be divided into four distinct phases: (i) planning; (ii) temporarily islanded; (iii) operation; and (iv) validation. The order of the design phases is crucial to ensure the proposed design meets the main goals of the use case(s). Moreover, if any objective function or constraint fails to converge, the planning scenario must be modified to handle the divergence. Here, operation optimisation tries to test the suggested solution in the planning and temporarily islanded stages under varying conditions and scenarios, employing either historical data or a large number of random scenarios (e.g., time-series Monte Carlo simulation).

5.7.1 Optimised Investment Capacities and Dispatch Planning Framework

This stage aims to optimise the selection of DER's type, location, and capacity based on the maximising the AW (minimising the costs) subject to the basic, voltage, thermal and single-phase and three-phase constraints. The optimisation will be conduct using Ipopt solver and using the microgrid components models (network, DER, and load). In this stage, a snapshot simulation will be carried out based on the best- and worst-case scenarios.

- **Scenarios**. In the planning stage, two main scenarios (best- and worst-case) are used in the snapshot simulation as follows:
  - Best-case scenario. This includes two parts: (i) high demand, with no non-dispatchable DERs; and low demand, with high non-dispatchable DERs generation.
  - Worst-case scenario. This includes two parts: (i) high demand, with high non-dispatchable DERs generation; and low demand, with non-dispatchable DERs generation.

- **Objective function**. The objective function aims to maximise the AW (i.e., minimise the investment and energy costs), when medium demand and medium non-dispatchable DERs generations. This includes the following parts:
  - DER, electrical storage battery and EVSE costs (e.g., initial unit investment costs, O&M costs, replacement costs, and fuel costs (for reciprocating generators)).
  - Service life and degradation factors of DERs, electrical storage battery and EVSE.
  - General economic parameters (e.g., discount rate, inflation rate, and electricity price escalation rates).
  - Annual demand growth rate.
  - Network tariffs and retailer energy rates to estimate earnings from the customers.
  - FiT rate.
  - Penalties on power factor, power quality, peak demand and curtailment if existed.
- **Constraints**. Several constraints should be set to ensure the technical requirements of the network and the DERs in planning microgrids. Those constraints can be listed as:
  - Voltage limits (minimum and maximum voltage limits).
  - Thermal limits (cables, transformers, and inverters maximum ampacity).
  - Dispatchable DERs constraints (maximum capacities).
  - If electrical storage battery is used, the energy discharge cannot be more than one hour of the energy capacity.
  - Non-dispatchable DERs cannot be dispatched.
  - The sum of power capacity of DDERs must be more or equal to the o sum of loads maximum (optional) (just for island-microgrid).
- **Verification**. If the proposed design is not meeting the requirements in this stage, then the algorithm will try other scenarios automatically. Otherwise, it will go to the temporarily islanded stage.

5.7.2 Microgrid Temporarily Islanded Framework

This stage aims to check if the islanded microgrid is possible based on the proposed design resulted from the planning stage and meets all the technical constraints. In addition, this stage will define the island zones. It also can curtail some loads and non-dispatchable DERs.

- **Scenarios**. The temporarily islanded on the proposed microgrid design can be tested in three scenarios based on the reliability requirements from the designer and according to the budget of the investment. These scenarios are as follows:

  - 1 h per day islanded.
  - Continuous islanded for 24-h.
  - Able to work 3-day in row as a continues islanded in case of blackout.

- **Objective**. Even in this stage no investment will be made, but the optimisation will be also conducted based on Ipopt solver to maximise the number of customers (e.g., critical customers) in the islanded microgrid scenario.
- **Constraints**. To maximise number of customers in the islanded microgrid scenario, several network and DER constraints should be considered as follows:

  - Voltage limits (minimum and maximum voltage limits).
  - Thermal limits (cables, transformers, and inverters maximum ampacity).
  - Dispatchable DERs constraints (maximum capacities).
  - Electrical storage battery constraints (maximum capacity, and maximum discharge rate)
  - EVSE (maximum charge rate).
  - Non-dispatchable DERs cannot be dispatched.

- **Verification**. If the proposed design from the planning stage is not meeting the requirements of the temporarily islanded, then the algorithm will go back to the planning stage to try other scenarios automatically. Otherwise, it will go to the operation stage.

5.7.3 *Optimised Microgrid Operation Framework*

Define the operation of the DERs considering energy costs, technical constraints, and lifespan (optional but required for ESS). The operation stage will be simulated using a time-series Montecarlo simulation and optimised using Ipopt solver.

- **Scenarios**. In this stage, a time-series simulation will be carried out based on one year or more historical data (high-resolution data of 1-h variation or less) or using a Montecarlo simulation based on a large number of random scenarios. The main aim is to run the simulation of the proposed design that passed the temporarily islanded stage under different cases and conditions including the uncertainties. Here, the operation of the non-dispatchable DERs will be tested with the option to optimise the capacities of the dispatchable DERs.
- **Objective**. The energy costs will be optimised in this stage to ensure the optimal operation of the dispatchable and non-dispatchable DERs.
- **Constraints**. A set of networks and DERs constraints should be considered in the operation stage as follows:

  - Voltage limits (minimum and maximum voltage limits).
  - Thermal limits (cables, transformers, and inverters maximum ampacity).
  - ESS constraints.
  - Dispatchable DERs constraints (maximum capacities).

- Electrical storage battery constraints (maximum capacity, and maximum discharge rate).
- EVSE (maximum charge rate).
- Non-dispatchable DERs cannot be dispatched.

- **Verification**. A manual verification is needed to ensure that the proposed solution is good enough and meets all the requirements. If this is failed to meet the main goal(s) and requirements, then back to the planning stage to try other scenarios and then follow the sequence to pass the temporarily islanded stage and then the operating stage.

Summary

Throughout the design phase, microgrid planning and operation can be established and evaluated to ensure optimal component selection and operation during the analysis phase. This should be determined based on load/generation profiles, DER costs and service life, energy costs, and technical constraints (e.g., network constraints, DER constraints, etc.). Therefore, optimising the microgrid's architecture can be separated into four independent phases: (i) planning, (ii) temporarily islanded, (iii) operation, and (iv) validation. The sequence of design phases is vital to ensuring that the suggested design fits the use case's primary objectives (s). In addition, if any objective function or constraint fails to converge, the planning scenario must be altered to account for the divergence. Here, operation optimisation uses either historical data or a large number of random scenarios to test the suggested solution in the planning and temporarily islanded stages under varied conditions and scenarios (e.g., time-series Monte Carlo simulation).

5.8 Optimal Power Flow

Optimal power flow (OPF) is an essential base for several applications in power systems, especially for solving the optimisation problems in power systems. In general, the mathematical equations used to solve the optimisation problem include the follows:

- Known parameters and unknown variables
- Constraints link with the parameters and variables through mathematical expressions
- Single or multiple objective function

The main objective of the OPF solver is to minimise or maximise the single or multiple objectives function and is based on the constraints of the variables and the power flow equation [58]. For example, for any microgrid, a cost function could be defined. The cost function shows how much it costs to produce energy in the power plants. Different DER technologies can be existed in a microgrid, some generate electricity at a lower cost than others. The objective is to find the optimal combination of DERs that minimises the cost function and thus produces energy as cheaply as possible. Therefore, the main aim of the OPF is to solve the power flow equations based on the single or multiple objectives function, considering a set of constraints [51].

The selection of the optimisation technique can be done as a trade-off between the accuracy, simplicity, model assumptions, data availability and simulation burden. Each microgrid planning process has its own set of goals and limitations, but some planning problems can be recognised as common to all microgrids. Thus, the relevant algorithms can be used in different system configuration by changing the objective function(s), and constraints associated with their parameters and variables [71]. In this context, optimisation techniques can be classified as conventional and AI techniques.

5.8.1 Conventional Techniques

Conventional techniques refer to models using the mathematical representation of the optimisation problem. These techniques include linear programming (LP), non-linear programming (NLP), mixed integer programming (MIP), and interior-point techniques. LP is one of the most widely utilised conventional techniques. The objective function and constraints are both linear functions of variables [72]. To reduce the complexity of computation, it is normally approximate LP by applying piecewise linear approaches and sequential approximation techniques. As for microgrid with AC power flow analysis, the OPF problem would be divided into two sub-problems: active and reactive power. These two sub-problems can be tackled simultaneously or sequentially and solved alternatively or separately. The popular utilised algorithms for solving LP include graphical method, simplex method, inner point method, etc.

NLP is another useful technique in mathematical optimisation, which represents by non-linear objective functions and constraints. NLP encompasses both unconstrained and constrained programming. Many simple and effective methods for solving unconstrained programming have been developed and successfully applied in power system applications [73]. When it comes to constrained programming, a common solving step is first transforming the problem into unconstrained programming via formulating an augmented objective function accompanying with extra penalty or converting the original function into an augmented Lagrangian objective function, afterwards, solving this unconstrained programming problem via applying valid optimisation algorithms. Moreover, there is a particular form of NLP composed of quadratic objective functions and linear constraints, called quadratic programming (QP). It can be applied to solve optimisation problems, for instance, operation problems with a quadratic objective function, and linearised constraints by Taylor expansion.

A MIP is one specific technique in which certain of the decision variables are restricted to be integer values at optima after solution. The introduction of integer variables broadens the range of useful optimisation problems that planners can construct and solve significantly (e.g., microgrid planning problem with installation decision) [74]. Based on the application of LP and NP, the MIP can be further divided into mixed-integer linear programming (MILP) and mixed-integer non-linear programming (MINLP).

In general, there are several mature tools and commercial solvers used conventional techniques can be applied to solve the OPF in microgrids, such as Gurobi, Cplex, GAMS, MOSEK, PENSDP, etc. Furthermore, numerous solvers, such as SDPA, SDPT3, CSDP, SeDuMi, and others, are publicly available [75]. Here, a list of the open-access and commercial convex optimisation solvers have been embedded in Yalmip (MATLAB platform) [76] and Pyomo (Python platform) [77].

The conventional techniques require explicit models and formulations to provide the optimal solution. Although the optimisation techniques described above can solve OPF, they are still focusing on improving the solving performance such as calculation speed, convergence, and starting points. Conventional techniques could not readily deal with discrete problems. Furthermore, the solving process and result are not as desired with the increasing complexity of problems and may never solve highly complex problems. This is because these optimisation techniques are based on the concept of local search. To obtain the global optimum, these methods rely on convexity, so the relationship needs to be simplified to assure convexity. For example, the actual OPF is generally non-convex, therefore, several local optimal values can exist and need to be compared. On the other hand, current researchers and developers proposed to use AI techniques, which eliminate the major shortcomings of conventional techniques and can tackle very complicated problems in a reasonable or effective solving time.

5.8.2 *Artificial Intelligence Techniques*

AI techniques refer to a group of intelligent algorithms analogising the process of natural evolution on basis of Darwin's principles or seek out the best solution to mimic a particular natural phenomenon [78]. The AI techniques can be classified in terms of optimisation-based techniques as heuristic, and meta-heuristic techniques.

Here, the heuristic method aims to reach a particular result of a problem tested by considerable experimentation that represents a compromise between the accuracy of the solution and the computational cost and speed. They cannot guarantee the result as the best solution, but it will reduce the computational load compared to conventional techniques. Meta-heuristics techniques are a high-level approach to the solve optimisation problems by combining multiple heuristics to reach the optimal or near-optimal solution [79]. Typical AI techniques include evolutionary algorithm (EA), genetic algorithm (GA), particle swarm optimisation (PSO), and simulated algorithm (SA). These AI techniques have been successfully applied to various power system optimisation problems where global solutions are extremely difficult to estimate or even non-differentiable region exist.

EA operates with a population of candidate solutions to the problem, which is similar to biological evolution. It usually includes evolutionary programming (EP). Being a sort of stochastic optimisation method, EA can theoretically reach global optimisation with a high probability. In OPF problems, EA is capable of handling discrete variables properly [78]. Also, EA has advantages in many aspects, including global convergence, concurrency, universality, robustness, etc. In addition, GA is one of the most widely used and well-known approaches. Based on the mechanisms of natural genetics, biological evolution and selection, this algorithm has demonstrated powerful capabilities and benefits in solving a variety of problems [80]. It is actually on basis of population evolution and selection, where the search process is performed by transforming one group of individuals in the search space into another. Each individual in the original GA is represented by a fixed-length binary string that maps a search space individual to an instance of an artificial chromosome. By altering the length of the binary string, the required accuracy can be approximated [81].

Moreover, PSO is one of the most important paradigms of swarm intelligence. It uses a simple approach that simulates the behaviour of a flock of birds or fish to help the particles in choosing the global optimal solution. Due to its ease of implementation, PSO has evolved rapidly in in latest decades, and it has many successful applications in tackling microgrid optimisation problems [82]. Further, SA is an iterative stochastic search process. it enables simulation on basis of molten metal's physical annealing mechanism. SA's operational philosophy is not complex. The difference between SA and common search methods is that each iterative step in SA has a certain probability of accepting some non-improving solutions. The global optima can theoretically be attained after numerous trials, but this requires a significant amount of processing effort [83].

The drawback of AI techniques is their imperfect computation efficiency. However, with the development of AI techniques, there are also some advanced approaches to improve the original AI techniques performance, such as approaches is self-adaptive modification technique [84], etc. In addition,, other AI algorithms have been applied in solving OPF, such as adaptive modified firefly algorithm [85], θ-PSO algorithm [86], etc. Thus, the selection of the AI algorithm depends on the formulation of the optimisation objectives and the nature of the constraints (i.e., binary, or continuous) as well as based on trade-off between the simplicity, accuracy, and computational burden.

Summary

Most recent planned and commencing projects around the country and worldwide aim to cover as many techno-economic aspects as possible. Here, it is normally to conduct an economic objective optimisation subjecting to technical performance constraints or conduct a technical performance objective optimisation subjecting to economic constraints or employing multiple objective functions (e.g., combining economic and technical, or adding objective function from social aspect and environmental aspect). The objectives vary based on the use case(s) and usually cover aspects such as cost efficiency, low environmental impact, and high reliability. To ensure the feasibility of microgrids subject to meet the design requirements (e.g., stability, reliability, power quality and security) at high-level of availability and at minimum cost, the planning, temporary islanded and operation optimisation problems should be well formulated. Further, it is an option to co-optimise the planning, temporary islanded and operation problems together.

In addition, the selection criteria of the proper optimisation techniques can be based on trade-off between the simplicity, accuracy, and computational burden. Then an optimal power flow can be conducted to achieve the targeted objective(s) subject to a set of constrains. Here, the optimisation techniques are mostly built on the top of the power flow engines to be able to conduct the OPF in microgrids. Although the conventional techniques are widely used in the optimisation tools, which are used in several practical projects, they have several simplifications, which are arising from lack of critical visibility of the data and modelling assumptions leading to grossly incorrect results. In addition, these conventional techniques are required long computational burden, which may negatively affect the digitalisation of the microgrids. Therefore, a need for faster and more accurate techniques to be able to integrate with the digitalisation revolution. In this context, AI techniques are showing promising results in the academic research, which can be an added value to improve the results in planning, temporary islanding and operating the microgrids as well as response faster to the control, communicate and market signals.

5.9 Simulation and Optimisation Tools

Several tools are available in the market for modelling, optimisation and simulating the microgrids. These tools are mainly categorised as power flow, optimal power flow and optimisation tools. Most of these tools are able to model several microgrid components and perform balanced or unbalanced steady-state power flow simulation under fundamental frequency and harmonics cases. However, some tools can conduct electromechanical and electromagnetic transient power flow (i.e., dynamic power flow). Few tools are able to provide unbalanced optimal power flow, which they can represent the real behaviour of the microgrid components with solving the optimisation problems. In this context, some common commercial tools (e.g., Hybrid optimisation model for multiple energy resources (HOMER) is just able to optimise and manage the components in microgrids without considering the unbalanced power flow), which may negatively affect the accuracy of the results and reduce the reliability of the service in the real-time operation. The tools can be categorised into open-access and commercial. These tools are described and compared in this section.

5.9.1 Power Flow Tools

The power flow of electrical networks can be simulated using several power flow solvers or algorithms. For this purpose, several tools (i.e., software and packages) are used. These tools can be classified according to their capability to conduct steady-state or dynamic power flow. It is worthy to mention that some dynamic power flow tools are able to conduct just balanced steady-state power flow simulation, which do not reflect the negative sequence representation of the network and the other microgrid components.

5.9.1.1 Steady-State Power Flow Tools

Different tools have been developed to conduct the steady-state power flow for different objectives and with different resource levels. This subsection presents an overview of the commercial, and open-access tools that developed for assessing unbalanced distribution networks [51]. The commonly used commercial tools are DIgSILENT PowerFactory [87], PSS/Sincal [88–90] and CYMDIST [91]. However, the most used open-access tools are open distribution system simulator (OpenDSS) [92–94], GridLAB-D [92], PandaPower [95, 96] and open platform for energy networks (OPEN) [97]. Accordingly, Table 5.17 summarises the differences between the commonly used steady-state power flow tools.

To sum up, the commercial tools (i.e., DIgSILENT PowerFactory, PSS/Sincal, and CYMDIST) can be used to study the unbalanced distribution network design and control variations based on fixed and time-series data. In addition, the open-access tools (i.e., OpenDSS, GridLAB-D, and OPEN) are capable to model and simulate the unbalanced distribution network under several condition with high range of accuracy, and they are open to collaboration. Several control strategies are supported by the stead state power flow simulations, which are significantly contribute to the design and operation of the microgrids. In this context, DIgSILENT PowerFactory provides the distributed, decentralised, and centralised control strategies for inverter-based resources, shunts and tap changers, while OpenDSS provides distribution control strategy for inverter-based resources and supports the smart control functions of the inverters. However, PSS/Sincal, CYMDIST and GridLAB-D supports the centralised control strategies at the network level not including the control strategies for inverter-based resources. In contrast, PandaPower and OPEN tools do not support the voltage regulation and control strategies.

Table 5.17 Comparison of the commonly used steady-state power flow tools

Features		DIgSILENT PowerFactory	PSS/Sincal	CYMDIST	OpenDSS	GridLAB-D	PandaPower	OPEN
Basic features	Power flow analysis for radial network	✓	✓	✓	✓	✓	✓	✓
	Power flow analysis for mesh/loop network	✓	✓	✓	✓	✓	✓	✗
	Power flow analysis for system with unbalanced loads	✓	✓	✓	✓	✓	✓	✓
	Source unbalance for slack bus	✓	✓	✗	✓	✓	✓	✓
	Multiple generation sources	✓	✓	✓	✓	✓	✓	✓
	Three-phase power flow analysis	✓	✓	✓	✓	✓	✓	✓
	Single-phase power flow analysis	✓	✓	✓	✓	✓	✓	✓
	Choosing the type of power flow technique	✓	✓	✓	✓	✓	✓	✓
	Voltage regulation	✓	✓	✓	✓	✓	✗	✗
	PV voltage regulation support in power flow	✓	✓	✗	✓	✓	✗	✗
Available equipment models	Distribution lines or cables (π-equivalent)	✓	✓	✓	✓	✓	✓	✓
	ZIP-load	✗	✗	✓	✓	✓	✓	✓
	Generators	✓	✓	✓	✓	✓	✓	✓

(continued)

Table 5.17 (continued)

Features		DIgSILENT PowerFactory	PSS/Sincal	CYMDIST	OpenDSS	GridLAB-D	PandaPower	OPEN
	Three-phase transformers	✓	✓	✓	✓	✓	✓	✓
	Models of PV generator	✓	✓	✓	✓	✓	✗	✓
	Storage units	✓	✓	✗	✓	✓	✗	✗
	Smart inverter control	✓	✗	✗	✓	✓	✗	✗
	Capacitors	✓	✓	✓	✓	✓	✓	✓
	Switches	✓	✓	✓	✓	✓	✓	✗
	Over-current devices	✓		✓	✓	✓	✗	✗
	Motors	✓	✓	✓	✓	✓	✓	✗
	Preloaded geographical PV insolation data	✓	✗	✗	✗	✓	✗	✗
	Residential end-user models	✗	✗	✗	✗	✓	✗	✗

5.9.1.2 Dynamic Power Flow Tools

Many dynamic microgrid analyses are conducted using root mean square (RMS) modelling and simulations. RMS only considers the fundamental frequency values, while EMT simulation considers the instantaneous value of voltage and current. Therefore, EMT simulations can be used to simulate very high-frequency phenomena (e.g., lightning or switching surges) [98]. RMS simulation tools can only model balanced systems (i.e., positive sequence) and are therefore unable to model individual phase values and accurately simulate unbalanced conditions. Inverters and wind turbines use state-of-the-art control strategies that require the measurement of voltages in individual phases or phases of positive/negative/zero sequences. Thus, a balanced, positive sequence RMS platform cannot always simulate modern inverter response correctly [98].

EMT simulators have evolved beyond being problem-solving tools for small and localised systems with the development of renewable energy systems and their complex controllers. EMT simulations have become an essential part of the design of large electric grids and microgrids [99]. EMT and RMS models will both always respond slightly differently from reality. The EMT and RMS models cannot be judged as more or less accurate without thorough validation. EMT models often consider higher frequency phenomena, so other than in a laboratory setting, they are challenging to validate. Positive sequence simulators are computationally efficient and afford efficient simulation of large power networks. However, in such a simulation environment, there can be concern regarding the accuracy of representation of inverter-based resources dynamic behaviour. The commonly used tools for dynamic power flow analysis are commercial tools, including power systems computer-aided design (PSCAD) [100] and PSS/E [101]. Even though DIgSILENT PowerFactory was covered within the steady-state power flow tools in the previous subsection, it also can be a part of dynamic tools as it supports the EMT and RMS domain simulation. Therefore, this tool is also included in the comparison with other dynamic power flow tools. Accordingly, Table 5.18 summarises the differences between the dynamic power flow tools in terms of basic features and a model library for most power components.

To summarise, PSCAD, PSS/E, and PowerFactory are the most used positive sequence simulation tools. as the default dynamic simulation environment by transmission planners for the past couple of decades. All the above-mentioned dynamic power flow tools can conduct EMT simulation, while just PSS/E and PowerFactory are capable to conduct RMS stability simulation. The main challenge for all these tools is the limitation in conducting full three-phase RMS models with positive/negative/zero sequences within short simulation time steps. These simulators were however designed from the perspective of synchronous machine electromechanical dynamics. Therefore, there can be concern regarding the accuracy of the representation of inverter-based resources dynamic behaviour using such tools.

Table 5.18 Comparison of the commonly used dynamic power flow tools

Features		PSCAD	PSS/E	DIgSILENT PowerFactory
Basic features	EMT simulation	✓	✓	✓
	RMS stability simulation	✗	✓	✓
	RTDS hardware-in-the-loop interface	✗	✗	✗
	Dynamic power flow analysis for system with unbalanced loads	✗	✗	✗
	Multiple generation sources	✓	✓	✓
	Choosing the type of power flow technique	✗	✓	✓
	Voltage regulation	✓	✓	✓
	PV voltage regulation support in dynamic power flow	✓	✓	✓
Available equipment models	Distribution lines or cables (π-equivalent)	✓	✓	✓
	ZIP-load	✗	✗	✗
	Dynamic load	✓	✓	✓
	Generators	✓	✓	✓
	Three-phase transformers	✓	✓	✓
	Models of PV generator	✓	✓	✓
	Storage units	✓	✓	✓
	Smart Inverter Control	✓	✗	✓
	Capacitors	✓	✓	✓
	Switches	✓	✓	✓
	Over-current devices	✓	✓	✓
	Motors	✓	✓	✓
	Preloaded geographical weather data	✗	✗	✗
	Residential end-user models	✗	✗	✗

5.9.2 *Optimal Power Flow Tools*

This subsection contains the most commonly used tools for the OPF problem in an unbalanced network. Numerous open access and commercial tools are available to solve the optimisation problem. The open access tools can be modified to improve the results. The reliability of the results can be increased by comparing the results of the open access tool with the well-known standardised tools. Furthermore, the used codes of most open-access tools are available for the public, which helps the developers to reproduce the results to review the performance of the tools. Two open-access tools are commonly utilised in the literature to solve the non-linear optimisation problem in an unbalanced distribution network, since the accessible commercial tools cannot solve the cannot solve non-linear optimisation problems in unbalanced distribution networks [51]. These tools are open-access, including Open-DSOPF [102] and PowerModelsDistribution [103]. Accordingly, the detailed comparison between Open-DSOPF and PowerModelsDistribution is given in Table 5.19.

To sum up, PowerModelsDistribution and Open-DSOPF are the open-access tools that support solving the non-linear optimisation problems in microgrids as well as the can model DERs-based AC grids. In addition, Open-DSOPF tool can implement the active management strategies to manage the reactive power and grid voltage level through the controllable devices in the network (e.g., inverter-based resources) in coordination with the OpenDSS tool. The support functions provided by the inverter behaviour control are volt/watt, volt/var, fixed power factor and other high-level functions to develop the intelligent inverter control characteristics. Accordingly, Open-DSOPF can just optimise the reactive power control in the network to reduce the active power curtailment. This optimisation function is not enough to design and operate the microgrids. In comparison, PowerModelsDistribution is not supporting the active management strategies for inverter-based resources at its current version, but it has the capability to be developed to handle several optimisation problems. Therefore, optimising the design and operation of microgrids require a modification of the objective function in Open-DSOPF, or developing an optimisation model using PowerModelsDistribution tool.

5.9.3 Other Tools

Several tools are used in the literature as investment and/or operation decision support tools for designing and operating the microgrids in grid-connected and islanded modes. Such tools are using to optimise and/or simulate the technologies behind the meter. Even they are considering the grid-connected operation mode, but the detailed model of network/feeder is not represented. In addition, such tools do not support the power flow analysis simulation. Accordingly, commonly used investment and/or operation decision support tools are covered in this subsection. Here, the system advisor model (SAM), and microgrid design toolkit (MDT) [104] are considered as open-access tools, while HOMER [105, 106], PVsyst [107], and i-hybrid optimisation by genetic algorithm (iHOGA) [108] are commercial tools. The comparison between the commonly used investment and/or operation decision support tools is listed in Table 5.20.

To sum up, the above-mentioned tools can be used to develop and simulate the microgrids at the components level, not at the system level as they are not supporting the power flow analysis neither in the steady-state domain nor in the dynamic domain. Therefore, such tools can be powerful in calculating the benefit/cost analysis for the energy trading in grid-connected and islanded operation mode at the steady-state. However, the stability and reliability of microgrids, protection, control and advance EMS cannot achieve using the above-mentioned tools. HOMER, iHOGA, MDT

Table 5.19 Comparison between Open-DSOPF and PowerModelsDistribution tools

Features		Open-DSOPF	PowerModelsDistribution
Basic features	OPF analysis for radial network	✓	✓
	OPF analysis for mesh/loop network	✓	✓
	OPF analysis for system with unbalanced load	✓	✓
	Multiple generation sources	✓	✓
	Three phase OPF analysis	✓	✓
	Choosing the type of OPF solver	✓	✓
	Choosing the type of objective function	✗	✓
	Voltage regulation	✓	✓
	PV voltage regulation support in PF	✓	✓
Available equipment models	Distribution lines	✓	✓
	DC line	✓	✓
	ZIP-load	✓	✓
	Generators	✓	✓
	Three phase transformers	✓	✓
	Model of PV generator	✓	✓
	Storage units	✓	✓
	Smart inverter control	✓	✓
	Capacitor	✓	✓
	Switches	✓	✓
	Overcurrent devices	✓	✓
	Motors	✓	✓
	Preloaded geographical weather data	✓	✗
	Residential end-user model	✓	✗

Table 5.20 Comparison between the commonly used investment and/or operation decision support tools

Features		HOMER	SAM	iHOGA	MDT	PVsyst
Basic features	Power flow analysis (steady-state and dynamic)	✗	✗	✗	✗	✗
	Time-horizon (1 h or less)	✓	✓	✓	✓	✓
	Techno-economic optimisation	✓	✗	✓	✓	✓
	Simulation	✓	✓	✓	✓	✓
	Multi-year simulation	✓	✗	✓	✓	✓
	Investment decision support	✓	✓	✓	✓	✓
	Operation decision support	✓	✗	✓	✗	✓
	Sensitivity analysis	✓	✓	✓	✓	✗
	Emissions consideration	✓	✗	✗	✗	✗
	Social parameters	✗	✗	✗	✗	✗
Available equipment models	General grid model (import/export)	✓	✓	✓	✓	✓
	Detailed network/feeder model	✗	✗	✗	✓	✗
	Grid-connected mode	✓	✓	✓	✓	✓
	Islanded mode	✓	✓	✓	✓	✓
	Hydropower plants	✓	✗	✓	✗	✗
	Loads	✓	✓	✓	✓	✓
	Diesel generators	✓	✓	✓	✓	✓
	Wind turbine	✓	✓	✓	✓	✗
	PV system	✓	✓	✓	✓	✓
	BESSs	✓	✓	✓	✓	✓
	Hydrogen fuel cells	✓	✓	✓	✗	✗
	Detailed inverter-based resources	✗	✗	✓	✗	✓
	AC bus	✓	✓	✓	✓	✓
	DC bus	✓	✓	✓	✓	✓
	Local control	✗	✗	✓	✓	✗
	Pricing structure	✗	✗	✓	✓	✗
	Preloaded geographical weather data	✓	✓	✓	✓	✓
	Import/export data as .csv	✓	✓	✓	✓	✓

and PVsyst can support techno-economic optimisation for renewable-based hybrid power systems, either islanded or grid-connected modes based on limited options of objective functions and ignoring the social aspects. In addition, the above-mentioned tools, except HOMER do not consider the emission effect in their simulation. In summary, the HOMER, iHOGA, MDT and PVsyst have a good optimisation base with detailed cost/benefit analysis, while they do not support a power flow simulation. Therefore, they can be used as investment and/or operation decision support tools to provide rough results, which can be used to validate the economic analysis of microgrid design and operation. Thus, more advance tools are required for planning, temporary islanding and operating of the microgrids considering a combination of the technical, economic, reliability, environmental and social aspects.

Summary

Modelling and designing the microgrid components into a network (or a part of it) need an accurate modelling approach and power flow/estate estimation simulation engine to ensure the real representation of all the components. The unbalanced steady-state time-series simulations are essential to reflect the time-dependent aspects of demand/generation profile and controllable devices in microgrids. This enables a more accurate assessment of voltages and power flows regarding time variation to support realistic dynamic "boundary conditions" in microgrid. This steady-state solution can be verified from an EMT simulation setup, either using dynamic power flow tools and/or hardware emulation systems.

In Australia, the most widely used power flow/estate estimation commercial tools are Digsilent PowerFactory (steady-state and dynamic tool), PSS/Sincal (steady-state tool), PSS/E (dynamic tool) and PSCAD (dynamic tool), while the most used open-access tools are OpenDSS (steady-state tool) and PandaPower (steady-state tool). In addition, HOMER and SAM tools are widely used as investment and/or operation decision support tools. All of the above-mentioned tools are either do not support OPF to design and operate the microgrids or do not perform unbalanced power flow. Even Open-DSOPF supports OPF, but it is just optimising the reactive power control in the network, which is not enough to design and operate the microgrids as microgrid requires other types of techno-economic optimisation functions. Moreover, PowerModels-Distribution has the capability to be developed to handle several optimisation problems, but it is still not supporting microgrids at its current version. In this context, well-defined objective function(s) needed to ensure the optimal design and operation of the microgrids, which could be obtained either by modifying the Open-DSOPF to include such objective functions or by developing new package based on OpenDSS software. Therefore, a tool that combines the optimal power flow in steady-state and dynamic domain including the ability to support the investment and operation stages considering the social and environmental aspects is needed to achieve the main goals of any microgrid project (i.e., high reliability at minimum cost). This tool can be built on the top of the power flow tools that mentioned in this section using an open-access high-level programming language (e.g., python or Julia). Although it is understood that positive sequence simulators would not be able to represent all the dynamics associated with inverter resources, recent advances in the development of positive sequence models for inverters in low short circuit networks offer promise to transmission system planners and operators, which can be implemented as a co-simulation with the steady-state power flow simulation to verify and validate the stability of the models in the microgrids.

References

1. Burk C. Techno-economic modeling for new technology development | AIChE
2. Gurba LW, Lowe A (2009) Techno-economic modelling of the energy systems: development of Australian conditions for technology assessment. Energy Procedia 1:4387–4394. https://doi.org/10.1016/J.EGYPRO.2009.02.253
3. EPRI (2012) Guidebook for cost/benefit analysis of smart grid demonstration projects. Palo Alto, California
4. Baghaee HR, Mirsalim M, Gharehpetian GB, Talebi HA (2016) Reliability/cost-based multi-objective Pareto optimal design of stand-alone wind/PV/FC generation microgrid system. Energy 115:1022–1041. https://doi.org/10.1016/J.ENERGY.2016.09.007
5. Rao SS (2009) Engineering optimization: theory and practice, 4th edn
6. Mun H, Moon B, Park S, Yoon Y (2021) A study on the economic feasibility of stand-alone microgrid for carbon-free island in Korea. Energies 14. https://doi.org/10.3390/EN14071913
7. HOMER Pro—microgrid software for designing optimized hybrid microgrids
8. Chen B, Chen C, Wang J, Butler-Purry KL (2018) Multi-time step service restoration for advanced distribution systems and microgrids. IEEE Trans Smart Grid 9:6793–6805. https://doi.org/10.1109/TSG.2017.2723798
9. Ye Z, Chen C, Chen B, Wu K (2021) Resilient service restoration for unbalanced distribution systems with distributed energy resources by leveraging mobile generators. IEEE Trans Ind Inform 17:1386–1396. https://doi.org/10.1109/TII.2020.2976831
10. Cleveland O, Ali Ahmed Andrew R Thomas OH, Henning M (2018) Techno-economic feasibility analysis of a microgrid in downtown. Cleveland, Ohio
11. Maqbool U, Tyagi A, Tyagi VV, Kothari R (2020) Optimization of the renewable-energy-based micro-grid for rural electrification in northern region of India. Clean Technol Environ Policy 22:579–590. https://doi.org/10.1007/s10098-019-01801-1
12. Suk H, Hall J (2019) Integrating quality of life in sociotechnical design: a review of microgrid design tools and social indicators. Proc ASME Des Eng Tech Conf 2B-2019. https://doi.org/10.1115/DETC2019-98005
13. Wolsink M (2012) The research agenda on social acceptance of distributed generation in smart grids: renewable as common pool resources. Renew Sustain Energy Rev 16:822–835. https://doi.org/10.1016/J.RSER.2011.09.006
14. Mathew M, Hossain MS, Saha S et al (2022) Sizing approaches for solar photovoltaic-based microgrids: a comprehensive review. IET Energy Syst Integr 4:1–27. https://doi.org/10.1049/esi2.12048
15. Zhao B, Zhang X, Chen J (2012) Integrated microgrid laboratory system. IEEE Trans Power Syst 27:2175–2185. https://doi.org/10.1109/TPWRS.2012.2192140

16. Di Silvestre ML, Graditi G, Sanseverino ER (2013) A generalized framework for optimal sizing of distributed energy resources in micro-grids using an indicator-based swarm approach. IEEE Trans Ind Inform 10:152–162
17. Lin N, Zhou B, Wang X (2011) Optimal placement of distributed generators in, pp 4239–4242
18. Caire R, Retiere N, Morin E et al. Voltage management of distributed generation in distribution networks. In: 2003 IEEE power engineering society general meeting (IEEE Cat. No.03CH37491). IEEE, pp 282–287
19. Wasiak I, Pawelek R, Mienski R, Gburczyk P (2012) Using energy storage for energy management and load compensation in LV microgrids. In: 2012 IEEE 15th international conference on harmonics and quality of power. IEEE, pp 904–908
20. Nguyen TA, Crow ML (2012) Optimization in energy and power management for renewable-diesel microgrids using dynamic programming algorithm. In: 2012 IEEE international conference on cyber technology in automation, control, and intelligent systems (CYBER). IEEE, pp 11–16
21. Nguyen DT, Le LB (2013) Optimal energy management for cooperative microgrids with renewable energy resources. In: 2013 IEEE international conference on smart grid communications (SmartGridComm), pp 678–683
22. Pregelj A, Begovic M, Rohatgi A, Novosel D (2003) On optimization of reliability of distributed generation-enhanced feeders. In: Proceedings of the 36th annual Hawaii international conference on system sciences, HICSS 2003. https://doi.org/10.1109/HICSS.2003.1173897
23. Abdelkader A, Rabeh A, Ali DM, Mohamed J (2018) Multi-objective genetic algorithm based sizing optimization of a stand-alone wind/PV power supply system with enhanced battery/supercapacitor hybrid energy storage. Energy 163:351–363
24. Sarhan A, Hizam H, Mariun N, Ya'acob ME (2019) An improved numerical optimization algorithm for sizing and configuration of standalone photo-voltaic system components in Yemen. Renew Energy 134:1434–1446
25. Bukar AL, Tan CW, Lau KY (2019) Optimal sizing of an autonomous photovoltaic/wind/battery/diesel generator microgrid using grasshopper optimization algorithm. Sol Energy 188:685–696
26. Tina G, Gagliano S, Raiti S (2006) Hybrid solar/wind power system probabilistic modelling for long-term performance assessment. Sol Energy 80:578–588
27. Nogueira CEC, Vidotto ML, Niedzialkoski RK et al (2014) Sizing and simulation of a photovoltaic-wind energy system using batteries, applied for a small rural property located in the south of Brazil. Renew Sustain Energy Rev 29:151–157
28. Alvaro D, Arranz R, Aguado JA (2019) Sizing and operation of hybrid energy storage systems to perform ramp-rate control in PV power plants. Int J Electr Power Energy Syst 107:589–596
29. Zhao B, Shi Y, Dong X et al (2013) Short-term operation scheduling in renewable-powered microgrids: a duality-based approach. IEEE Trans Sustain Energy 5:209–217
30. Lamedica R, Santini E, Ruvio A et al (2018) A MILP methodology to optimize sizing of PV-Wind renewable energy systems. Energy 165:385–398
31. Corso G, Di Silvestre ML, Ippolito MG et al (2010) Multi-objective long term optimal dispatch of distributed energy resources in micro-grids. In: 45th international universities power engineering conference UPEC2010. IEEE, pp 1–5
32. Duan L, Zhang R (2013) Dynamic contract to regulate energy management in microgrids. In: 2013 IEEE international conference on smart grid communications, SmartGridComm 2013, pp 660–665. https://doi.org/10.1109/SmartGridComm.2013.6688034
33. Chen C, Duan S, Cai T et al (2011) Smart energy management system for optimal microgrid economic operation. IET Renew Power Gener 5:258–267
34. Modes S, Jiang Q, Xue M et al (2013) Energy management of microgrid. 28:3380–3389
35. García F (2013) Optimal economic dispatch for renewable energy microgrids with hybrid storage using model predictive control, pp 7932–7937
36. García F (2013) Regulation service for the short-term management of renewable energy microgrids with hybrid storage using model predictive control, pp 7962–7967

37. Adika CO, Wang L (2014) Autonomous appliance scheduling for household energy management. 5:673–682
38. Ma K (2013) Energy consumption scheduling in smart grid: a non-cooperative game approach
39. Adika CO, Wang L (2013) Energy management for a customer owned grid-tied photovoltaic micro generator
40. Ng SKK, Member S, Zhong J, Member S (2012) Smart dispatch of controllable loads with high penetration of renewables, pp 1–6
41. Zhang Y, Gatsis N, Giannakis GB (2013) Robust energy management for microgrids with high-penetration renewables. IEEE Trans Sustain Energy 4:944–953. https://doi.org/10.1109/TSTE.2013.2255135
42. Erol-Kantarci M, Kantarci B, Mouftah HT (2011) Cost-aware smart microgrid network design for a sustainable smart grid. In: 2011 IEEE GLOBECOM workshops (GC Wkshps). IEEE, pp 1178–1182
43. Urias MEG, Sanchez EN, Ricalde LJ (2014) Electrical microgrid optimization via a new recurrent neural network. IEEE Syst J 9:945–953
44. Ajlan A, Tan CW, Abdilahi AM (2017) Assessment of environmental and economic perspectives for renewable-based hybrid power system in Yemen. Renew Sustain Energy Rev 75:559–570
45. Adefarati T, Bansal RC (2019) Reliability, economic and environmental analysis of a microgrid system in the presence of renewable energy resources. Appl Energy 236:1089–1114
46. Chaouachi A, Kamel RM, Andoulsi R, Nagasaka K (2013) Multiobjective intelligent energy management for a microgrid. IEEE Trans Ind Electron 60:1688–1699. https://doi.org/10.1109/TIE.2012.2188873
47. Hirsch A, Parag Y, Guerrero J (2018) Microgrids: a review of technologies, key drivers, and outstanding issues. Renew Sustain Energy Rev 90:402–411. https://doi.org/10.1016/j.rser.2018.03.040
48. EPRI (2012) Stochastic analysis to determine feeder hosting capacity for distributed solar PV. EPRI Tech Updat 1026640:1–50
49. El-Bidairi KS, Nguyen HD, Jayasinghe SDG et al (2018) A hybrid energy management and battery size optimization for standalone microgrids: a case study for Flinders Island, Australia. Energy Convers Manag 175:192–212
50. Upadhyay S, Sharma MP (2014) A review on configurations, control and sizing methodologies of hybrid energy systems. Renew Sustain Energy Rev 38:47–63
51. Ardakani FJ, Riahy G, Abedi M (2010) Optimal sizing of a grid-connected hybrid system for north-west of Iran-case study. In: 2010 9th international conference on environment and electrical engineering. IEEE, pp 29–32
52. Alsayed M, Cacciato M, Scarcella G, Scelba G (2013) Multicriteria optimal sizing of photovoltaic-wind turbine grid connected systems. IEEE Trans energy Convers 28:370–379
53. Sinha S, Chandel SS (2015) Review of recent trends in optimization techniques for solar photovoltaic–wind based hybrid energy systems. Renew Sustain Energy Rev 50:755–769
54. Shrestha GB, Goel L (1998) A study on optimal sizing of stand-alone photovoltaic stations. IEEE Trans Energy Convers 13:373–378
55. Maghraby HAM, Shwehdi MH, Al-Bassam GK (2002) Probabilistic assessment of photovoltaic (PV) generation systems. IEEE Trans Power Syst 17:205–208
56. Ogunjuyigbe ASO, Ayodele TR, Akinola OA (2016) Optimal allocation and sizing of PV/Wind/Split-diesel/Battery hybrid energy system for minimizing life cycle cost, carbon emission and dump energy of remote residential building. Appl Energy 171:153–171
57. Ma G, Xu G, Chen Y, Ju R (2017) Multi-objective optimal configuration method for a standalone wind–solar–battery hybrid power system. IET Renew Power Gener 11:194–202
58. Paliwal P, Patidar NP, Nema RK (2014) Determination of reliability constrained optimal resource mix for an autonomous hybrid power system using particle swarm optimization. Renew Energy 63:194–204
59. Calvillo CF, Sánchez-Miralles A, Villar J (2015) Assessing low voltage network constraints in distributed energy resources planning. Energy 84:783–793. https://doi.org/10.1016/J.ENERGY.2015.03.044

60. Petrelli M, Fioriti D, Berizzi A et al (2021) A novel multi-objective method with online Pareto pruning for multi-year optimization of rural microgrids. Appl Energy 299:117283. https://doi.org/10.1016/j.apenergy.2021.117283
61. Anderson K, Farthing A, Elgqvist E, Warren A (2022) Looking beyond bill savings to equity in renewable energy microgrid deployment. Renew Energy Focus 41:15–32. https://doi.org/10.1016/J.REF.2022.02.001
62. Wächter A, Biegler LT (2006) On the implementation of an interior-point filter line-search algorithm for large-scale nonlinear programming. Math Program 106:25–57. https://doi.org/10.1007/s10107-004-0559-y
63. Forsgren A, Gill PE, Wright MH (2002) Interior methods for nonlinear optimization. SIAM Rev 44:525–597. https://doi.org/10.1137/S0036144502414942
64. Gould N, Orban D, Toint P (2005) Numerical methods for large-scale nonlinear optimization. Acta Numer 14:299–361. https://doi.org/10.1017/S0962492904000248
65. Capitanescu F, Martinez Ramos JL, Panciatici P et al (2011) State-of-the-art, challenges, and future trends in security constrained optimal power flow. Electr Power Syst Res 81:1731–1741. https://doi.org/10.1016/j.epsr.2011.04.003
66. Wächter A (2017) Chapter 17: Nonlinear optimization algorithms. Advances and trends in optimization with engineering applications. Society for Industrial and Applied Mathematics, Philadelphia, PA, pp 221–235
67. Wächter A (2002) An interior point algorithm for large-scale nonlinear optimization with applications in process engineering. PhD thesis
68. Wächter A, Biegler LT (2005) Line search filter methods for nonlinear programming: local convergence. SIAM J Optim 16:32–48. https://doi.org/10.1137/S1052623403426544
69. Wächter A, Biegler LT (2005) Line search filter methods for nonlinear programming: motivation and global convergence. SIAM J Optim 16:1–31. https://doi.org/10.1137/S1052623403426556
70. Nocedal J, Wächter A, Waltz RA (2009) Adaptive barrier update strategies for nonlinear interior methods. SIAM J Optim 19:1674–1693. https://doi.org/10.1137/060649513
71. Australian Energy Market Operator (2020) Renewable integration study Appendix A: high penetrations of distributed solar PV important notice, pp 1–54
72. Mo N, Zou ZY, Chan KW, Pong TYG (2007) Transient stability constrained optimal power flow using particle swarm optimisation. IET Gener Transm Distrib 1:476–483. https://doi.org/10.1049/IET-GTD:20060273
73. Nagrath IJ, Kothari DP, Desai RC (2008) Modern power system analysis
74. Kleinert T, Labbé M, Ljubić I, Schmidt M (2021) A survey on mixed-integer programming techniques in bilevel optimization. EURO J Comput Optim 9:100007. https://doi.org/10.1016/J.EJCO.2021.100007
75. Löfberg J (2004) YALMIP: a toolbox for modeling and optimization in MATLAB. In: Proceedings of the IEEE international symposium on computer-aided control system design, pp 284–289. https://doi.org/10.1109/cacsd.2004.1393890
76. YALMIP
77. Pyomo
78. Yilmaz L, Tolk A (2008) Intelligent decision making: an AI-based approach 97:193–226. https://doi.org/10.1007/978-3-540-76829-6
79. Makhadmeh SN, Khader AT, Al-Betar MA et al (2019) Optimization methods for power scheduling problems in smart home: survey. Renew Sustain Energy Rev 115:109362. https://doi.org/10.1016/j.rser.2019.109362
80. Holland JH (1992) Adaptation in natural and artificial systems: an introductory analysis with applications to biology, control, and artificial intelligence. Adapt Nat Artif Syst. https://doi.org/10.7551/MITPRESS/1090.001.0001
81. Fogel DB (2000) Applying Fogel and Burgin's 'competitive goal-seeking through evolutionary programming' to coordination, trust, and bargaining games. In: Proceedings of 1997 IEEE international conference on evolutionary computation, ICEC, vol 2, pp 1210–1216. https://doi.org/10.1109/CEC.2000.870788

82. Cai M, Zhang X, Tian G, Liu J (2007) Particle swarm optimization system algorithm. In: Advanced intelligent computing theories and applications with aspects of contemporary intelligent computing techniques, pp 388–395. https://doi.org/10.1007/978-3-540-74282-1_44
83. Hsiao YT, Liu CC, Chiang HD, Chen YL (1993) A new approach for optimal VAR sources planning in large scale electric power systems. IEEE Trans Power Syst 8:988–996. https://doi.org/10.1109/59.260902
84. Rigo-Mariani R, Sareni B, Roboam X, Turpin C (2014) Optimal power dispatching strategies in smart-microgrids with storage. Renew Sustain Energy Rev 40:649–658. https://doi.org/10.1016/J.RSER.2014.07.138
85. Khan B, Singh P (2017) Selecting a meta-heuristic technique for smart micro-grid optimization problem: a comprehensive analysis. IEEE Access 5:13951–13977. https://doi.org/10.1109/ACCESS.2017.2728683
86. Baziar A, Kavousi-Fard A (2013) Considering uncertainty in the optimal energy management of renewable micro-grids including storage devices. Renew Energy 59:158–166. https://doi.org/10.1016/J.RENENE.2013.03.026
87. DIgSILENT GmbH (2021) PowerFactory. https://www.digsilent.de/en/powerfactory.html. Accessed 16 Apr 2021
88. Siemens AG (2021) PSS®SINCAL. https://new.siemens.com/global/en/products/energy/energy-automation-and-smart-grid/pss-software/pss-sincal.html#:~:text=PSS®SINCALoffers avarietyofmodulesforthe,tacklethedifferentchallengesofthechanginggrid. Accessed 15 Apr 2021
89. Siemens AG (2018) PSS®SINCAL: all-in-one simulation software for the analysis and planning of power networks. Erlangen, Germany
90. Siemens AG (2021) PSS®SINCAL platform electricity modules. https://new.siemens.com/global/en/products/energy/energy-automation-and-smart-grid/pss-software/pss-sincal/pss-sincal-electricity.html. Accessed 15 Apr 2021
91. EATON (2020) CYMDIST. https://www.cyme.com/software/cymdist/. Accessed 13 Apr 2021
92. Chassin DP, Schneider K, Gerkensmeyer C (2008) GridLAB-D: an open-source power systems modeling and simulation environment. In: 2008 IEEE/PES transmission and distribution conference and exposition. IEEE, pp 1–5
93. Rizy DT, Staunton RH (2002) Evaluation of distribution analysis software for DER applications. Oak Ridge, TN
94. Dugan RC (2010) Open distribution simulations system workshop: using open DSS for smart distribution simulations. In: EPRI PQ smart distribution 2010 conference and exhibition, pp 14–17
95. Thurner L, Scheidler A, Schafer F et al (2018) Pandapower—an open-source python tool for convenient modeling, analysis, and optimization of electric power systems. IEEE Trans Power Syst 33:6510–6521. https://doi.org/10.1109/TPWRS.2018.2829021
96. PandaPower. http://www.pandapower.org/
97. Energy and Power Group—University of Oxford (2020) OPEN. https://github.com/EPGOxford/OPEN. Accessed 14 Apr 2021
98. PowerFactory Di (2017) Power system engineering and software
99. Transients E PSCAD V5
100. PSCAD (2014) PSCAD Users Guide V4_6_0
101. Siemens AG (2017) PSS/E-high-performance transmission planning and operations software for the power industry
102. Rigoni V, Keane A (2020) Open-DSOPF. https://github.com/ValentinRigoni/Open-DSOPF. Accessed 26 Apr 2020
103. Fobes DM, Claeys S, Geth F, Coffrin C (2020) PowerModelsDistribution.jl: an open-source framework for exploring distribution power flow formulations. Electr Power Syst Res 189:106664. https://doi.org/10.1016/j.epsr.2020.106664

104. Eddy J (2017) Microgrid design toolkit (MDT) user guide software v1.2. Albuquerque, NM, and Livermore, CA
105. HOMER Pro. www.homerenergy.com
106. Homer Energy (2020) Robust hybrid microgrid optimization modelling, pp 1–2
107. PVsyst photovoltaic software. www.pvsyst.com
108. IEEE (2021) IEEE standard for dc microgrids for rural and remote electricity access applications. IIEEE Std 2030.10-2021 1–47

Chapter 6
Microgrid Integration and Interactions with the Main Grid

6.1 General

Integration of DERs and loads is one of the main challenges of microgrids. Modes of operation of the microgrid, transitions between modes, steady-state, and dynamic characteristics of the microgrid, protection schemes, and control system architecture need to be well understood. System integration and operation requires dynamic and transient studies as well as detailed dynamic and short circuit simulations [1]. In this section, the essential aspects of microgrid integration and interactions with the main grid are briefly described.

6.2 Microgrid Characteristics and Distribution Energy Resources Operation

Compared to a bulk power system, microgrid characteristics are different; microgrids are smaller, have lower inertia, limited short circuit capacity, higher R/X ratio, unbalanced loading, and have DERs that are power electronically interfaced. The characteristics of multiple DER and their controllers, as well as the load, should be taken into account before integrating them in a microgrid (e.g., low short circuit capacity of the DER, the high inrush current of motor loads, and system inertia in microgrids in islanded mode). In addition, the operation of a microgrid under a significant unbalanced condition can cause system instability, so proper techniques must be implemented to handle these situations. For instance, a power oscillation may occur if the voltage unbalances is not properly compensated [1].

DERs can act as either grid-forming or grid-following units in a microgrid. The term "grid-forming unit" refers to assets that have the ability to set voltage and frequency (e.g., diesel or gas generators in the isochronous mode or droop mode, and BESS, including inverters with grid forming capability). The grid-following DERs

I. A. Ibrahim et al., *Microgrids Design and Operation*, Power Systems,
https://doi.org/10.1007/978-981-96-4914-3_6

are generators or ESSs that synchronise with online assets and follow the grid voltage and frequency. These normally operate in PQ (i.e., specified active and reactive power setpoints) mode or MPPT mode. Some DER can function in both modes, and the DER's local controller or microgrid control system should be able to shift from one mode to the other. Note that each DER's primary controller has a different characteristic for each mode of operation, and if the modes of operation are not handled properly through the microgrid controller, voltage or frequency instability can be occurred [1].

6.2.1 Disconnection Devices

During a grid fault/failure, an appropriate disconnection device should be in place in order to separate and reconnect the grid by receiving the appropriate commands from the control system. Occasionally, the microgrid's existing switches/breakers are unable to meet this need. In addition, the disconnection time of the switches may not be fast enough to mitigate or suppress voltage sags/swells and maintain system stability in case of a grid failure. In real-world microgrid deployments, some changes are required, such as replacing incoming disconnect switches and/or load break switches with motor-operated switches and adding fast-acting static switches [1].

6.2.2 Ungrounded Medium Voltage System

Ungrounding refers to the absence of intentional grounding on a power system. Due to the very low fault current level in the ungrounded systems, there is a minimal equipment damage. In spite of this, there is always the possibility of equipment damage resulting from transient overvoltage in ungrounded systems in microgrids due to the arcing after a fault or switching at a breaker if the insulation levels for the equipment are inadequate [2].

Voltage is the most accurate way to determine if a ground fault has occurred because the current is very low. A zig-zag grounding transformer [1] or wye-grounded–broken-delta voltage transformer [2] is considered as effective methods of protecting the microgrid against ground faults by providing a path for the fault current. As far as continuity of service is concerned, ungrounded systems do not offer any benefits over grounded systems with high resistance and are prone to transient over voltages and difficulty in finding ground faults [3].

6.2.3 Synchronisation to a Weak Grid

When a weak grid hosts the microgrid, the synchronisation process and control system should be appropriately handled through the synchronisation controller and microgrid control system, and the inner loop controllers of the DER need to be tuned properly [1]. The Canadian Renewable Energy Laboratory (CanREL) has demonstrated a synchronisation process involving a 200 kW battery system and a 100 kW wind turbine simulator connected to a weak grid [4, 5]. Due to the incorrect tuning of the inner loop controllers, there was a slight phase deviation during synchronisation, which caused the inverter of the wind turbine simulator to trip.

6.2.4 Transformer Energisation

Due to the large amount of magnetising current required for the bank of transformers, the start of the backup unit/units in islanding mode may fail. The solution would be to disconnect some of the transformers, preferably using remote-controlled devices. To avoid inrush currents, a soft magnetisation process provided by the backup generator excitation could be used to energise the bank of transformers. Here, the use of inverter-based DER to power large transformers necessitates the use of certain methodologies and procedures. Because inverter-based DERs have a lower maximum current contribution capability than synchronous machine-based DERs, inverter-based grid-forming units must be properly scaled, and approaches must be adopted. To manage inrush currents, several strategies such as adding automation energisation at transformers with variable frequency drive (VFD), or energisation procedure through inverter-based DER can be used to energise transformers through the inverter before synchronisation to the grid [1].

6.2.5 Voltage and Frequency Control

As a result of the high R/X ratio, frequency and voltage controls are not fully decoupled in microgrids. The microgrid control system and DER primary controllers are responsible for controlling voltage and frequency on islanded microgrids. In the absence of properly controlled voltage and frequency by grid-forming DERs and the microgrid controller, sudden changes in generation or load may cause voltage and frequency instability. The control of inverters in islanded low-inertia microgrids and frequency control are discussed in many studies [6]. However, the most common microgrid deployments use simple frequency and voltage droop control [1].

6.2.6 *Grid Integration*

Both grid-connected and islanded microgrids need to meet certain quality, reliability, and protection standards, making their integration into the utility crucial. Controllable DERs and loads need to be actively involved to keep the system voltage and frequency stable. In the context of grid-connected microgrid operations, the whole microgrid can be considered as one controllable entity. Some Canadian utilities (e.g., Hydro One, Ontario, Canada) apply distributed generation interconnection requirements for the interconnection of microgrid with the grid, based on IEEE 1547.4 guidelines. For integrating inverter-based resources, some utilities are using North American Electric Reliability Corporation (NERC) guidelines [1]. Like those in data centres, a microgrid connected to the grid can provide grid services like frequency and voltage control to the main grid. A grid-interactive uninterruptible power supply (UPS) is one example of DER providing frequency regulation by controlling the charging and discharging of batteries.

Summary
Microgrids have the potential to become a key building block for future smart grids due to better reliability, stability, and higher power quality. In islanded operation mode, microgrids can face control and protection issues due to the infrequent behaviour of DERs and non-linear loads. In addition, adequate voltage and frequency regulation are essential to avoid the instability caused by an abrupt load change. By disconnecting some transformers and non-critical loads, smart switches can support the stable operation of a microgrid. Further, microgrids may encounter utility integration challenges in grid-connected operation mode, leading to control and protection issues. The right power sharing and utility integration method is crucial for the stable and controlled operation of a microgrid. Therefore, controllable DERs and loads need to be actively involved to keep the system voltage and frequency stable in both operation modes. In the literature review, researchers are looking for feasible solution to address the integration challenges using several strategies, including droop control strategies, tuning of the inner loop controllers, and trials of grid-forming inverter-based DERs.

6.3 Microgrid Roles on Grid Resiliency

The increasing penetration of DERs makes microgrids a viable solution to control and scalability problems. A microgrid could also be a solution for supplying power to critical/sensitive loads during low-frequency, high-impact events, improving overall system resiliency. Catastrophic events like hurricanes, earthquakes, and wildfires frequently cause extended power outages, sometimes lasting days or weeks. In such circumstances, the potential to reconfigure parts of the distribution system as microgrids could provide local communities with critical services, such as providing power to local hospitals, temporary shelters, and community centres, while the rest of the power system is repaired. Microgrids can play an important role in resiliency in these situations. Another new development emerging in research and development with some demonstration projects is merging multiple microgrids to create a networked microgrid [1].

6.3.1 Modelling for Resiliency

Microgrids must provide a stable platform for operations to support resiliency. During normal and expected abnormal events, it must be able to maintain a stable frequency and voltage. In grid-connected operation mode, sufficient energy is required to supply the load and cover the losses. In some cases, the ability to lose the single largest generator when islanded without an interruption to critical end-use loads [1]. Microgrids must be able to maintain system stability while operating during routine operations in addition to steady-state considerations. In complex transitions between grid-connected and islanded operations after power outages, microgrids may be operating in conditions different from those studied during design. Therefore, various forms of stability analysis (e.g., power supply, balance, and control system stability) become even more critical, which should be considered during the design stage [7]. These conditions can be tested either using stability simulation tools or using real-time emulations. Microgrid stability behaviours can differ from large and interconnected bulk power systems. Accordingly, several requirements must be considered in the designing stage to ensure the stability and reliability of the microgrids as follows [1]:

- System voltage stability as well as DC link stability for inverter-based resources.
- Unintentional islanding, where short-term frequency stability could become relevant.
- Small disturbance stability related to either balance stability and power supply (voltage and frequency stability) or to control system stability (electric machinery and converter stability).
- Large load starting, including in-rush, could result in short-term frequency stability and long-term voltage stability.
- Fault response that could result in long-term electric machine and converter stability and short-term voltage stability.

To evaluate the capability of the microgrid to operate in a stable and reliable environment, the following types of simulations must be conducted [1]:

- Unbalanced steady-state time-series power flow.
- Electromagnetic dynamics, considering network dynamics.

- Electromechanical dynamics with unbalanced network (ignores network dynamics).
- Techno-economic optimisation (production cost model type studies).
- Protection analysis.

6.3.2 Benefits of Resiliency

The value of microgrids' ability to provide energy during long-term outages caused by natural disasters such as hurricanes, earthquakes, wildfires, etc., is difficult to quantify. An extended electricity outage can have a significant impact; however, such situations do not often occur, making their impact harder to assess [1].

Communities' societal burdens have been measured through risk-based approaches, which describe how much effort is required to obtain basic and critical services [8, 9]. These evaluations can identify microgrids' resiliency value to communities. Incorporating these evaluations into common industry practices still requires significant work.

Summary

Microgrids have been developed to help mitigate the reliability and stability issues as well as address the effects of natural disasters and to prevent wildfires. Therefore, microgrids should be able to supply the energy to the loads (e.g., critical/sensitive loads) during the outages and low-frequency events, boosting overall system resiliency. Accordingly, microgrid must be able to maintain system stability, while operating in all the modes, especially in islanded operation mode. Therefore, system voltage stability, unintentional islanding, small disturbance stability, large load starting, and fault response must be examined by appropriate types of simulations to ensure the capability of the microgrid to operate in all the conditions.

6.4 Microgrid Control

To ensure efficient operation, microgrids usually are equipped with different controllers (e.g., primary, secondary, and tertiary control layers). For example, the synchronisations of DERs in the microgrid and between the microgrid and main grid as well as the energy balancing and management can be determined by the microgrid controller. According to the interconnection regulations, the DERs without controllers in a microgrid are capable solely of connecting and disconnecting, which can negatively affect the microgrid's operation and reliability. Thus, the controllers are the brain of the microgrid, acting as a decision-making firmware/software and hardware [10]. The implementation of the microgrid control can be fully decentralised, distributed across a network of intelligent devices or centralised in a single microgrid control unit called microgrid central controller. Accordingly, the microgrid controller can contribute to [1]:

- Providing ancillary services to the grid and the exchange of power.
- Transiting between grid-connected and islanded modes.
- Controlling voltage and regulate frequency.
- Optimising the operation of local assets.
- Deciding on load curtailment and shifting as well as DER and charging/discharging curtailment.

Hence, the microgrid control functions can be classified into different groups based on their technical and economic requirements as follows [1]:

- **Balancing supply and demand in short-term**. The function of the primary control is to balance demand and supply in short-term, which ensures frequency and voltage stability of microgrids during disturbances (e.g., temporary islanding). The other functions are the compensation of power quality, and regulation of load shedding schemes.
- **Managing energy in long-term**. The main objective of tertiary control (e.g., long-term energy management tool) is to dispatch the controllable resources of the microgrid (e.g., DERs, ESSs, EVSE and controllable loads), and control the power flow between the microgrid and the main grid. The business model and operation modes of microgrids decide management objectives. During the islanded mode of a microgrid, the energy management tool ensures sufficient reserve capacity to secure its operation.
- **Tools for assessing security**. The microgrid security tool aims to identify its unsecured operating point, which can cause unstable operation in the event of unplanned islanding. During islanded operation mode, a load shedding scheme is implemented to supply power to critical/sensitive loads if the microgrid does not have sufficient local generation capacity.
- **Restoration process for microgrids**. The restoration process of microgrid after the blackout or islanding failure is initiated by the central controller. Local service restoration is accomplished using a fully automatic method that coordinates the operation of local resources. The primary control will stabilise the system.

The core level functions of the microgrid controller are between functions that are device specific within the microgrid and distribution grid specific across the PCC with the distribution grid, which can be summarised as shown in Fig. 6.1.

In general, the microgrid controller has two main functions as follows [1]:

- Core function 1—Dispatch
  - Maximises asset utilisation, including DER, smart inverter-based resources, and conventional sources.

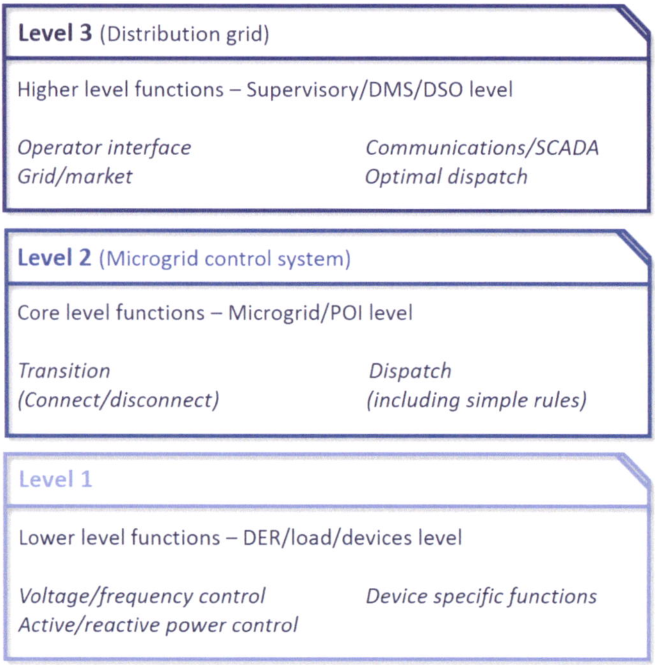

Fig. 6.1 Microgrid controls system functional framework and core functions

- Ensures that the microgrid's operation satisfies the standards and requirements both internally and from the PCC to the distribution system.
- During steady-state operation, executes commands at regular intervals (usually 15 min or fewer).
- During transitions, executes emergency dispatch orders to ensure that transitions are completed in a timely and orderly manner, dispatching assets, and controlling the load.

• Core function 2—Transition

In the following cases, manages transitions and operates:

- Unplanned islanding due to a loss of utility grid.
- Planned islanding because of the DSO's request to disconnect.
- Reconnection/resynchronisation.
- Black start as part of the process of establishing an island.
- Switches the dispatch function from one mode to another by sending a control signal.

The controller's functions are the same for all microgrids independent of topology or configuration. The functional requirements are set by the microgrid operator (internal and grid interface) and the DSO (interconnection). The most common control architecture for microgrids is a 3-layer hierarchical structure comprised of

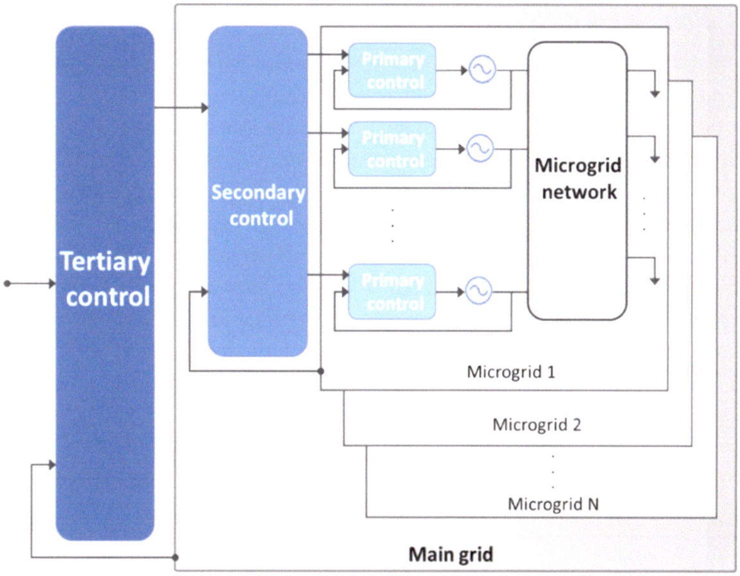

Fig. 6.2 Microgrid control hierarchical structure

primary, secondary and tertiary controls, as depicted in Fig. 6.2. While this control structure is drawn from the traditional 3-level temporal decomposition of frequency control in bulk power systems, the hierarchical structure adopted for microgrids also considers distinctive features.

The primary or local control features the fastest response and relies only on local measurements. It operates in a timeframe of seconds or less. Given their speed requirements, control tasks included in the primary level do not rely on communication networks. The energy management is carried out through secondary control. This control level includes the restoration of voltage and frequency to their desired set-points, and the economic and reliable operation of the microgrid in both, grid-connected or islanded operation modes, as well as the transition between them. Secondary control appears to be the highest hierarchical level in islanded microgrids and operates on a slower time frame compared to the primary level, to decouple their dynamics and allow for more sophisticated optimisation-based controls. There are different ways to perform energy management or secondary control ranging from centralised to semi- or fully distributed. At the highest level of control, the tertiary control level is responsible for coordinating the microgrid operation with the main grid and the neighbouring microgrids as well. In this level, control objectives include the voltage and frequency support of the host grid, active and reactive power dispatch settings at the PCC for loss minimisation or congestion control, and power exchange between microgrids. The operation of this controller is in the order of several of minutes and it transfers signals to secondary controller of microgrids [11].

6.4.1 Primary Control Level

Microgrid islanding detection, DER output controls (voltage control and current regulation), and power-sharing/balance management are all considered as local control. The primary control level is considered as local controllers. While the output control and power-sharing tasks in synchronous generators are handled by the conventional automatic voltage regulator (AVR), governor, and the machine's inertia; droop controllers can be used to share power of inverter-interfaced DERs [1]. Primary voltage and frequency regulation are crucial for the microgrid's stability in an emergency mode. To ensure this, controllers must respond rapidly and robustly. Therefore, these functions are implemented locally in VSIs coupled with ESSs through external control loops. The primary control level has the fastest response and only uses local measurements. It works in a fraction of a second or less [1].

6.4.1.1 DER Control

The local control of the DERs can be categorised into two types based on the operation mode of microgrids as follows [12]:

- **Local control during the grid-connected mode**. The local control parameters in grid-connected mode are injected active and reactive power at the PCC. For smooth interconnection, the inverter's voltage magnitude and phase angle must match those at the interconnection point. The point of connection of DERs with the grid includes a filter, an inverter, and a pulse width modulation (PWM) controller. The inverter controller regulates the output voltage of the inverter so that DER output voltage is always in synchronisation with the PCC.
- **Local control during the islanded mode**. In the islanded mode, at least one DER should be control under the grid-forming mode to maintain magnitude and frequency of the voltage. Here, the inverter controller is responsible for regulating voltage magnitude and phase at the interconnection point and ensuring the DER is compatible with the rest of the system. The inverter settings would be set to always follow the reference voltage and frequency in a single inverter-based DER system.

6.4.1.2 Load Control

Load control strategies are designed based on the type of load. In households, for example, loads are divided into three categories, namely critical/sensitive, interruptible, and deferrable [12].

- **Critical/sensitive loads** are those that cannot be shed or shut down even in an emergency situation.

- **Controllable loads** are those that are capable of shifting to later hours during shortages or high energy prices (e.g., water heating, clothes drying, and other plug-in loads).
- **Interruptible loads** are defined as those that can be curtailed in times of emergencies or high electricity prices. Examples include space cooling and optional lighting.

There are four inputs required by the load control model: (i) price signal or emergency signal from an external source; (ii) the load profile; (iii) percentage of allowable demand reduction (e.g., 0% for critical/sensitive loads and 100% for interruptible loads); and (iv) percentage of compensation after deferring the load.

6.4.2 Secondary Control Level

The secondary control level entails restoring voltage and frequency to their target setpoints and the microgrid's economical and reliable operation in both grid-connected and islanded operation modes, as well as the transition between them. Furthermore, the energy management system can be performed in the secondary control level. In islanded microgrids, it is the highest hierarchical level and works at a slower rate than the primary level. Accordingly, the controllability of dispatchable micro-sources could also be used to perform secondary frequency regulation, which would reduce the power requirements of storage units performing primary frequency regulation [1].

6.4.3 Tertiary Control Level

Tertiary control is the highest level of control hierarchy in microgrids. It coordinates the operation of a microgrid and its host grid, as well as multiple nearby microgrids; therefore, it can be considered part of the host grid. Control objectives belonging to the tertiary level can include the voltage and frequency support of the host grid, active and reactive power dispatch settings at the PCC for loss minimisation or congestion control, and power exchange between microgrids. The tertiary control level usually operates within a few minutes, providing signals for secondary level controls at microgrids [1].

6.4.4 Control Methods Based on Communication Links

Furthermore, based on the communication links used in the microgrid, the control strategies of the microgrids can be further categorised into centralised, distributed, and decentralised.

6.4.4.1 Centralised

In a centralised architecture, a central controller collects all necessary data to make dispatch decisions, which are then delivered to the primary controls of each controllable asset. The central controller communicates and supervises each local controller to ensure stable operations in both grid-connected and islanded modes [13]. In this architecture, it is possible to achieve highly detailed optimisation models that can be solved centrally. Generally, centralised architecture is considered more appropriate for microgrids which all controllable assets are managed by the microgrid operator [1]. Moreover, accurate power sharing algorithms can be implemented in this control level in order to avoid overloading of certain inverters in the microgrids, which could endanger the whole microgrid reliability as it may triggering protection relays and causing cascaded failure in the microgrid [14].

6.4.4.2 Distributed

There is no central controller in a fully distributed secondary control architecture; instead, each controllable DER possesses have some level of intelligence and communicates with one another through a bi-directional communication network to accomplish local and global secondary objectives. An example of this approach is the multi-agent-based microgrids. In addition to providing a plug-and-play feature, fully distributed approaches are more suitable for microgrids with multiple DER owners [1].

6.4.4.3 Decentralised

In this type of controller architecture, the inverter-based resources are controlled locally based on its output measurements. In other words, there is no secondary control layer as well as there is not any communication between neighbour inverter-based resources. This architecture can manage controlling the DER units to handle some critical functions (e.g., reliability, flexibility, re-dispatch after sudden isolation, other emergency situations, communication disruption, and a plug-and-play feature) [1].

Summary
Microgrids required several levels of control based on the application and complexity. For a given microgrid at least should have the primary control level, as it is the basic control level that directly controls the output power of the inverter-based resources. The primary control level is considered as the fastest response type. While the secondary control is the additional control level that is mainly used to restore the voltage and frequency of the microgrid at the load point of connection to the desired values. Additionally, the secondary control can be used for energy management purposes. While the primary and secondary control levels are utilised to control the microgrids in islanded operation mode, the tertiary control level is used to enable the microgrid connection with the main grid for the purpose of optimal economic dispatch (i.e., export/import power between the microgrid and main grid). Based on the reviewed national and international projects, the primary and secondary are the most used control levels in microgrids. In addition, the proper selection of the control technique and its settings play an important role to ensure the stability of the microgrid. To do so, the exact performance requirements and services should be defined based on the standards. It is recommended to take into account the flexibility in the performance and the operation of the control architecture of the inverters. Otherwise, it may lead to complex design and/or instability of the microgrid.

6.5 Microgrid Protection

Microgrids can increase the percentage of electricity availability to consumers by providing power in the event of a utility grid disruption or outage. Protection of the microgrid is required when a short circuit occurs within the confines of the microgrid to protect the flexibility of the microgrid [15]. The availability of protection devices for distribution systems is not a suitable solution for microgrid protection as the short circuit current capacity of these devices is very high. The difference between the fault current level of a microgrid and that of a distribution system is significant, and this difference makes coordination difficulties [16].

Accordingly, numerous methods and technologies have been proposed by researchers in the literature to protect the microgrid, but these methods require extensive testing and verification before they can be considered commercially viable. The only way to protect the microgrid from the extreme challenges of variable fault current levels and high costs is using transmission and distribution system protection devices (e.g., distance, overcurrent, differential relays, etc.). The characteristics of synchronous generators are used to design transmission and distribution system protection devices. Since the characteristics of DERs are different from those of a synchronous generator, it is necessary to modify or upgrade the above-mentioned protection devices to consider the unique characteristics of DER and ensure the reliable and safe operation of the microgrid [17].

DC microgrids, unlike AC microgrids, are commonly used in low, medium, and high voltage DC applications. Protecting DC microgrids is more complicated and difficult than protecting AC microgrids, and these difficulties limit the adoption of DC microgrids [18].

In microgrid, there are two types of faults that can occur as follows:

- **External faults**. This fault comes from the grid feeder to the microgrid in grid-connected mode. This fault occurs outside the boundary of the microgrid. The external fault disrupts the synchronisation between the grid feed-in and the microgrid and causes unintentional islanding [19].
- **Internal faults**. An internal fault can occur in grid-connected, or islanded mode and this fault occurs within the boundary of the microgrid. When this fault occurs, the microgrid central controller redefines the outer boundary of the microgrid using intelligent switches and an optimisation algorithm to eliminate the effect of the fault on the system [13].

6.5.1 Fundamental Requirements of Microgrid Protection

The basic microgrid protection requirements that are considered when designing a protection device for a microgrid are listed as follows [20]:

- **Selectivity**. The ability of a protection system to isolate only the faulty part of a microgrid while the rest of the system continues to function.
- **Stability**. The ability of a protection system to prevent remote faults.
- **Speed of Operation**. The ability of a protection system to operate as quickly as possible to reduce the risk of damaging the microgrid assets.
- **Sensitivity**. An additional measure of protection against the hazards of contact with live parts.
- **Primary Protection**. This is the primary safeguard for any microgrid circuit.
- **Backup Protection**. This is a backup protection system that kicks in if the primary protection system fails to response for any fault.

6.5.2 Contribution of Different Micro-sources to Fault Current

DERs contain synchronous and asynchronous generators as well as inverse-based resources. Therefore, the DERs can contribute to the fault current in both grid-connected and islanded operation modes. The contribution of the fault current from different sources is examined and explained in detail below [21].

6.5.2.1 Contribution of Synchronous Generators to Fault Current

The fault current of the synchronous generators decreases exponentially until it reaches a stationary level. During a fault, the generator terminal voltage drops dramatically and the field current drops proportionally. Since a fault current last only a fraction of a second, a moulded case circuit breaker (MCCB) cannot detect it. A field override mechanism is required to maintain field current in the event of a fault so that the MCCB can detect and recover from the fault [22].

6.5.2.2 Contribution of Asynchronous Generators to Fault Current

Asynchronous generators are used in small-scale hydro power plants to generate electricity. In the event of a fault, the terminal voltage of the generator is drastically reduced and there is also no air gap flux. A fault current is initially high and then quickly decreases [23].

6.5.2.3 Contribution of Inverter-Based Resources to Fault Current

Most of the DERs are connected to the microgrid via power converters. The current regulation mechanism is part of the inner control loop of the power converters. In the event of a fault, the internal control circuit of the power converters regulates the positive sequence system of its output current to a constant value in the time between the response of the main protection system and the islanding detection. Insulated gate bipolar transistors (IGBTs) are used in the power converters; therefore, a high current flow for a short period of time can damage them. To protect the IGBT from damage, a control mechanism is implemented to shut off the DERs in case of high transients. Compared to other conventional sources, the contribution of the fault current from sources connected to transformers is very small [24].

6.5.3 Protection of Islanded Microgrids

The protection of an islanded microgrids is discussed at length when the microgrid system is not connected to the utility grid. The protection requirements and equipment in islanded operation mode are discussed in this subsection.

6.5.3.1 Protection for Safety

During the development phase of the microgrid, the protection strategy of the microgrid for people or animals living nearby is crucial. To protect against electric shock, basic protection such as isolation of microgrid equipment, barriers and enclosures can be used to prevent anyone from touching and reaching the microgrid [25].

6.5.3.2 Overcurrent and Short Circuit Protection

Overcurrent and short circuit protection devices are designed to disconnect equipment or conductors in the event of a fault. The most used protection devices to protect the microgrids from the overcurrent and short circuit events are coved in this subsection.

- **Typical protective devices**. Several protection equipment can be used in microgrids. Here, the most used protection equipment are listed as follows [26]:
  - **Fuses**. A fuse is essentially a thin wire connected in series with the power line or other equipment. A thin wire heats up and melts when a fault current flows through it, breaking the conductor. The current continues to flow in the form of an arc after the conductor is broken, and quenching powder is used in the fuse to eliminate the arc. The current rating and the maximum current of the fuse should be determined based on short circuit analysis in the microgrid.
  - **Reclosers**. It is a protection device for fault identification and repair. It includes an intelligent control unit that can distinguish between temporary and permanent faults. Reclosers are often used at the PCC of a microgrid. In the event of a fault state, it disconnects the circuit and then reconnects it after a period of time to ensure that the problem is still present. The operation will be repeated three to four times by the recloser until fixing the fault.
  - **Miniature circuit breakers (MCBs)**. This safety device consists of a bimetallic strip and a magnetic coil. It can interrupt both long-term overcurrent and short circuit currents. In the event of a short circuit current, a magnetic coil breaks the circuit, while in the event of a sustained overcurrent condition, a bimetallic strip flexes and breaks the circuit [26].
  - **Residual current devices**. It is used to quickly disconnect the circuit in the event of a ground fault. It consists of two magnetic coils, one connected to the phase and the other to the neutral wire. Under normal conditions, the same amount of current flows through both coils, but in the event of a ground fault, the current in both coils differs, resulting in a net magnetic flux in the coil. The net magnetic flux generates electromotive force to trip the circuit breaker.
  - **MCCBs**. They work in the same way as MCBs, with the exception that MCCBs have a higher rating and breaking capacity than MCBs. Type-B MCCBs are used for power factor protection, Type-C MCCBs are used for inductive loads protection, and Type-D MCCBs are used for motor loads protection.
  - **Disconnector**. It is a protection device that can disconnect a line conductor manually or it can be programmed or controlled from the control stations. It is used to disconnect a switch in case of maintenance or permanent fault condition.

- **Protection of solar plants**. Fuses and disconnectors, which can be connected to each string of a PV array to protect it in the event of a short circuit current or overcurrent condition, can be used to protect solar power plants in the microgrid. Because fuses are exposed to temperature and radiation, the fuse's capacity and rating can be calculated to account for radiation and the maximum ambient temperature [27]. The same protection mechanism can be utilised to protect ESSs and EVSE.
- **Protection of small-scale hydropower and wind power plants**. The protection devices MCB and MCCB are commonly used in small-scale hydropower and wind power plants to protect them from overcurrent and short circuit current [28].
- **Protection of diesel/synchronous generators**. MCCBs are commonly used to protect diesel/synchronous generators against overcurrent and short circuit current. A field override method is required to handle the fault current detection and recovery [29].

6.5.3.3 Surge Protection

Surge arresters or surge protection devices (SPDs) can be used to protect the microgrid from lightning by switching surges by diverting the surge to earth instead of traveling through the conductors and damaging the equipment [30].

Summary

The complexity of the microgrid architecture influences its protection when connected to the grid. To protect the microgrid against internal faults of micro-sources interfaced with transformers, differential protection systems can be used. For overcurrent protection within and outside the microgrid network, the necessary coordination of overcurrent protection devices should be considered to ensure the safety of the components. In under and over frequency conditions, rate of change of frequency relays can be used to disconnect the microgrid from the utility grid. However, if the fault occurs outside the microgrid, the micro-sources must rid-through the fault in accordance with standards.

In design stage of microgrids, residual current circuit devices should be considered for the small-scale DERs to protect the people and cattle that live nearby. In addition, MCCBs or MCBs should be to protect the microgrid in overcurrent and short circuit condition. While SPDs must be taken into consideration to protect the microgrid against high voltage transients caused by lightning. For the appropriate operation of microgrid, a proper attention must be given to the earthing requirement, which should be maintained according to the requirements mentioned in the protection standards used by DNSPs.

6.6 Energy Management and Real-Time Operation

The balance between demand and supply needs to be maintained in microgrids in both grid-connected and islanded operation modes with satisfying certain reliability, power quality, and adequacy standards. To achieve these objectives, all DERs and loads should be directly or indirectly involved, depending on their control capabilities, in keeping the system voltage and frequency stable and within acceptable ranges. Localisation of energy production and consumption has created opportunities for new energy management technologies, which have already shown the potential to lower energy costs and improve energy supply reliability and security [31].

There are several technical and economic advantages of employing small-size DERs in the low-voltage distribution networks. Microgrids make it possible to manage an energy system systemically. The monitoring and management of microgrid assets are conducted by an energy management system (EMS). An EMS can be implemented as a hybrid system consisting of software and hardware that provides a decision-support platform for microgrids' operators [32]. The EMS in microgrids can provide various functionalities, including:

- Monitoring of the microgrid under various conditions.
- Capacity to handle a wide range of threats.
- Analysing different operational states of the system.
- Taking control actions.
- Determining the exchange of energy with the main grid.
- Making quick decisions during critical situations.
- Handling the interactions with different agents.

The general structure of EMS in microgrid is indicated in Fig. 6.3. The following subsections will analyse its different tasks.

6.6.1 Monitoring System

The EMS monitoring system offers an accurate and thorough view of the continuous operation of the microgrid. Meters and sensors gather this information and transmit it to EMS through the remote terminal units (RTUs) by means of the communication infrastructure. The monitoring system can provide data on [32]:

- Real-time status of switches, lines, and transformers.
- Power output from DERs.
- Output power and SOC of ESSs and EVSE.
- Information of the reactors and capacitors.
- Fuel availability of dispatchable DERs.

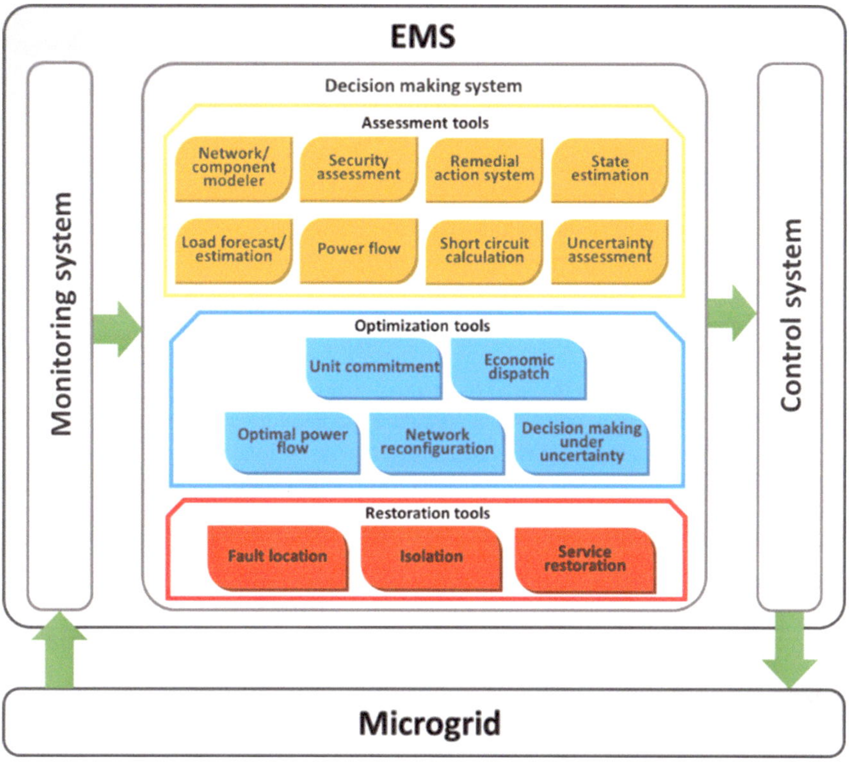

Fig. 6.3 General structure of EMS in microgrids

6.6.2 Control System

EMS control strategies depend on the complexity and importance of the microgrid procedures. Sometimes, EMS controls the components directly during automatic control. A self-repairing microgrid, for example, automatically identifies the fault, isolates it, and restores the system. Supervisory control is another option in which the system operator can make direct changes to the status of switches through the EMS when needed [32].

6.6.3 Decision-Making System

The decision-making feature enables the EMS to assess, optimise, and restore the operation of microgrids. In the subsequent subsections, the decision-making feature are discussed.

6.6.3.1 Assessment Features

The assessment tools enable the microgrid operator to check the ongoing microgrid operating conditions using assessment tools. The following is a list of these tools [32].

- **Microgrid device model**. This tool includes the mathematical model for microgrid components' behaviour and characteristics.
- **State estimation**. Using the few metered data, this tool can estimate the state variables of microgrids.
- **Load forecast**. This tool predicts load demand in a different time frame (e.g., hourly, daily, monthly, yearly), using forecasted weather conditions, and historical data. In addition, it uses the load patterns of different load groups.
- **Power flow**. System operators can calculate the network's power flows and node voltages using this tool in different configurations or load levels.
- **Fault current**. It calculates the currents during short circuit and compares them to their allowable values.
- **Uncertainty evaluation**. This tool analyses uncertain parameters and develop models, including their impacts in system performance.
- **Security computation**. This tool enables the microgrid operator to identify the most critical contingencies, such as violation of frequency range due to reduced system inertia.
- **Remedial action scheduling**. A system operator could use this tool to handle system contingencies. It offers a variety of solutions such as generation unit start-up, system reconfiguration using programs (e.g., emergency demand response program (EDRP)).

6.6.3.2 Optimisation Features

As different load-supply strategies exist in microgrids, they require a systematic approach to making decisions. The optimisation tool should provide a wide range of applications based on the microgrid operation strategy and relevant economic and technical factors, outlined as follows [32]:

- **Unit commitment (UC)**. This tool determines the availability of generation units to meet the load demand, taking into account the economic objectives and technical properties of DERs.
- **Economic dispatch**. This optimisation module determines the optimal output power of active DERs in microgrids.
- **OPF**. The OPF module can have different objectives, including cost minimisation, loss minimisation, and profit maximisation.
- **Network reconfiguration**. This tool calculates the best power path for the loads from different perspectives with reduced losses and higher reliability. Reconfiguration can also be used to share the load across different lines.

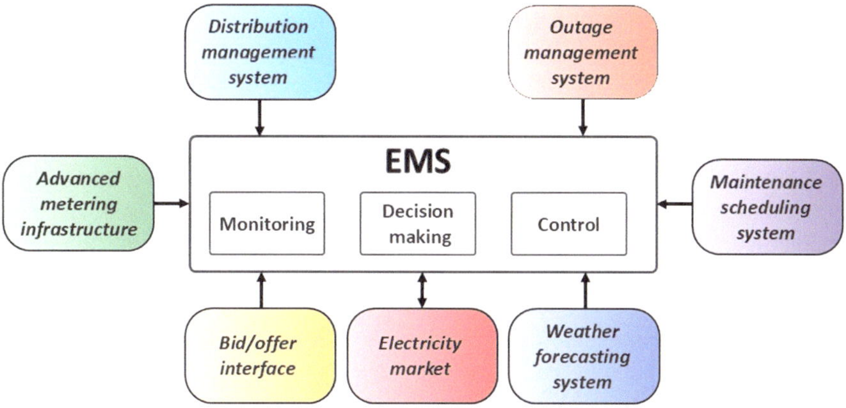

Fig. 6.4 Interaction among EMS and other systems

- **Decision making under uncertainty**. It provides different methods of decision-making based on uncertainty. The module includes robust optimisation, stochastic programming, information gap decision theory, and so on.

6.6.3.3 Restoration Features

The system operator can use this tool to restore the normal system operation after the occurrence of faults. The identification and isolation of faults can be performed automatically or manually [32]. Restoring the interrupted loads and black start procedure are also essential.

6.6.4 Interaction with Other Systems

Several functions of EMS are achieved through interaction with other systems as shown in Fig. 6.4. An EMS interacts with several systems, such as advanced metering infrastructure (AMI), distribution management system (DMS), outage management system (OMS), maintenance scheduling system, weather forecasting system, electricity market, and bid/offer interface [32].

6.6.5 Energy Management Systems Implementation

The implementation of an EMS can be done either in a centralised or decentralised manner. A computational core dispatches commands to local controllers to operate various components in centralised EMS, whereas local controllers are intelligent agents that analyse and dispatch their associated components based on the local conditions in decentralised EMS [32].

A centralised EMS applies optimisation procedures to efficiently manage the microgrid based on the different collected information. The main functions of a centralised EMS are economic dispatch and UC. By continually updating the input data, scheduling can reduce the effect of uncertainties. Several decision-making approaches are utilised to implement UC and economic dispatch (e.g., example machine learning systems, optimisation, and rule-based systems). For real-time applications, a centralised EMS is suitable as several microgrid components are monitored and analysed in one place [32].

Different agents schedule different parts of microgrid in decentralised energy management. The method works well on large-scale microgrids when the DERs and loads are distributed, and centralised data gathering is not viable or feasible. In decentralised decision-making, multi-agent systems (MAS) are an important tool for power management. Each agent manages an device which makes decisions based on its own data as well as data from other agents [32].

6.6.6 Hour-Ahead and Real-Time Energy Management System

As microgrids contain various energy sources, energy storage devices, etc., they have a complex structure and more data sources. Consequently, most of its energy management issues are considered to be offline optimisation issues related to typical hour-ahead scheduling, which requires early predicting of generation/demand profiles before optimising. However, as distributed energy is intermittent, and most loads can change instantaneously, it is hard to make predictions that are close to reality [33]. Currently, offline optimisation algorithms mainly include MIP, PSO, robust optimisation, and sequence QP algorithms.

Hour-ahead EMS are invalid for real-time scheduling due to the uncertain and intermittent operation of microgrids. Therefore, EMS must run an optimal scheduling problem as closely as possible to the real-time operation. Indeed, EMS requires a framework that applies hour-ahead scheduling over the hours of the operation horizon, for example, 24 h ahead, and then uses real-time scheduling for energy management over smaller time steps near the real-time operation time [32].

Real-time EMS strategies have been proposed to solve hour-ahead EMS drawbacks by dividing the timeline into small intervals to improve energy management efficiency [33]. Accordingly, the results are applied to the real-time EMS that is repeated in 5-min time steps. Even though real-time microgrid EMS schedules the operation for the next time steps at the end of current time-step, the system's next hour forecasted condition is indirectly considered by utilising the latest hour-ahead scheduling results [32]. Mostly, real-time EMS is approached using a model-based paradigm. As a general concept, the model-based approach consists of developing an explicit model of microgrid dynamics, a predictor of uncertainties, and an optimiser that solves the best schedules. A popular approach for real-time EMS is model predictive control (MPC). A major advantage of MPC is its ability to self-correct the forecast based on uncertainties of models and to self-adjust the sequence of controls using repeated optimisation of the predicted model over a finite time-horizon. The performance of the model-based controller may also degrade if accurate models or appropriate estimates of the parameters are unavailable [34].

Recently, machine and deep learning-based schemes have been proposed for studying the microgrid energy management issue. Machine learning techniques can handle uncertainty without the requirement of detailed microgrid models which is treated as a black box. However, most machine learning-based approaches suffer from the curse of dimensionality and struggle to handle microgrids with bigger-size state-variables and uncertainties. Therefore, deep learning techniques are proposed a few years ago to overcome the that challenges by utilising deep neural networks' capability to learn from end-to-end [34].

6.6.7 System Operation Strategies

EMS can employ various operation strategies, including microgrid maintenance and operational, cost of battery degradation, energy transaction, outages and interruption, losses cost, demand response incentives, load shedding penalty cost, emission cost, and levelised cost of renewable energy resources. There are some technical constraints in EMS, such as thermal cable limits, limit of renewable energy output power, energy balancing, reliability, reactive power, and other physical limits of assets [32]. Overall, the operation strategies are categorised as economic and technical strategies are summarised in Fig. 6.5 and discussed further in the following subsections.

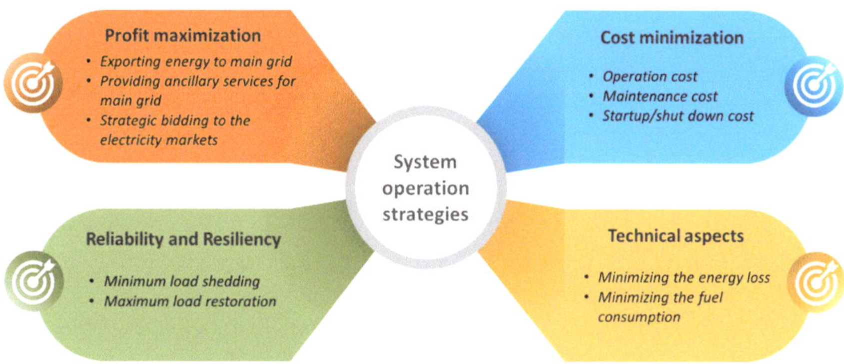

Fig. 6.5 System operation strategies of EMS

6.6.7.1 Economic Aspects

The main economic objective of EMS is to minimise the total operation cost or maximise the owner's profit. The grid-connected microgrid can export additional power to the main grid. Additionally, it can import power from the main grid to supply internal consumers. To meet the needs of consumers, the internal resources in microgrid compete with its upstream network. The cost of energy will be an essential input parameter of decision-making in this case, as it is best to make decisions with the highest profit margin. As microgrid EMS includes fast responsive generation and controllable loads, it can be profitable to utilise as back-up generation and reserve capacity for the main grid [35]. Moreover, the main important economic strategy of the islanded microgrids is cost minimisation [32].

6.6.7.2 Technical Aspects

The main objective of the technical aspect is to maximise system reliability indices and reduce energy losses or fuel consumption. Reduction of losses/fuel consumption is in line with cost reduction, but the enhancement of reliability can be conflicting with cost optimisation [32].

Summary

The EMS is a critical component of a microgrid that is responsible for monitoring and controlling all interactions and operating resources according to specified strategies. Maintaining the balance between demand and supply in grid-connected and islanded operation mode is the main goal while fulfilling certain standards regarding reliability, power quality, and adequacy. EMS offers a variety of analytical tools for analysing the microgrid operation; among them, the optimal operation tool is the most popular. According to the reviewed national and international projects, minimising costs of the energy and maximising the profits are the most popular operations strategies.

Due to microgrids' uncertain and intermittent operation, a two-stage scheduling procedure was introduced in the recent microgrid projects, incorporating the hour-ahead and real-time EMS. A unit commitment model is included in the hour-ahead scheduling and is run each hour to schedule the microgrid from the start of the operation horizon through the end. Using the economic dispatch model, the real-time EMS only takes into account the current hour while considering the on/off status of DERs from the latest results of hour-ahead scheduling. Recently, machine and deep learning-based schemes are being investigated to overcome the drawbacks of model-based approaches in EMS.

References

1. Sub-Committee on MicroGrids Whitepaper, IEEE Smart Grid Technical Activities Committee (2022) Microgrids: utility challenges and opportunities
2. Blackburn JL, Domin TJ (2006) Protective relaying: principles and applications. CRC Press
3. (2001) IEEE recommended practice for protection and coordination of industrial and commercial power systems (IEEE Buff Book). IEEE Std 242-2001 (Revision IEEE Std 242-1986) [IEEE Buff Book] 1–710
4. Arriaga M, Nasr E, Rutherford H (2017) Renewable energy microgrids in northern remote communities. IEEE Potentials 36:22–29
5. Wittemund M, Manjunath MR, Wrathall N. Canadian Renewable Energy Laboratory (CANREL)—a testbed for microgrids
6. Farrokhabadi M, Cañizares CA, Simpson-Porco JW et al (2019) Microgrid stability definitions, analysis, and examples. IEEE Trans Power Syst 35:13–29
7. Farrokhabadi M, Canizares CA, Simpson-Porco JW et al (2020) Microgrid stability definitions, analysis, and examples. IEEE Trans Power Syst 35:13–29. https://doi.org/10.1109/TPWRS.2019.2925703
8. Jeffers RF, Hightower MM, Brodsky NS et al (2017) A grid modernization approach for community resilience: application to New Orleans, LA. Sandia National Lab. (SNL-NM), Albuquerque, NM
9. Jeffers RF, Baca MJ, Wachtel A et al (2018) Analysis of microgrid locations benefitting community resilience for Puerto Rico. Sandia National Lab. (SNL-NM), Albuquerque, NM
10. Hatziargyriou N, Asano H, Iravani R, Marnay C (2007) Microgrids. IEEE Power Energy Mag 5:78–94. https://doi.org/10.1109/MPAE.2007.376583
11. Guerrero JM, Vasquez JC, Matas J et al (2011) Hierarchical control of droop-controlled AC and DC microgrids—a general approach toward standardization. IEEE Trans Ind Electron 58:158–172. https://doi.org/10.1109/TIE.2010.2066534
12. Rahman S, Pipattanasomporn M (2010) Modeling and simulation of a distributed generation-integrated intelligent microgrid. Arlington, VA
13. EPRI (2018) A smart and flexible microgrid with dynamic boundary and intelligent open-source controller. Palo Alto, California
14. Mohammed N, Lashab A, Ciobotaru M, Guerrero JM (2022) Accurate reactive power sharing strategy for droop-based islanded AC microgrids. IEEE Trans Ind Electron 1–1. https://doi.org/10.1109/TIE.2022.3167141
15. Sortomme E, Venkata SS, Mitra J (2010) Microgrid protection using communication-assisted digital relays. IEEE Trans Power Deliv 25:2789–2796. https://doi.org/10.1109/TPWRD.2009.2035810
16. Che L, Khodayar ME, Shahidehpour M (2014) Adaptive protection system for microgrids: protection practices of a functional microgrid system. IEEE Electrif Mag 2:66–80. https://doi.org/10.1109/MELE.2013.2297031
17. Sharaf HM, Zeineldin HH, El-Saadany E (2018) Protection coordination for microgrids with grid-connected and islanded capabilities using communication assisted dual setting directional overcurrent relays. IEEE Trans Smart Grid 9:143–151. https://doi.org/10.1109/TSG.2016.2546961
18. Zhang L, Tai N, Huang W et al (2018) A review on protection of DC microgrids. J Mod Power Syst Clean Energy 6:1113–1127. https://doi.org/10.1007/s40565-018-0381-9
19. Rebollal D, Carpintero-Rentería M, Santos-Martín D, Chinchilla M (2021) Microgrid and distributed energy resources standards and guidelines review: grid connection and operation technical requirements. Energies 14
20. Achparaki M, Thessalonikeos E, Tsoukali H et al (2012) We are IntechOpen, the world's leading publisher of open access books built by scientists, for scientists TOP 1 %. Intech 13
21. Emarati M, Barani M, Farahmand H et al (2021) A two-level over-voltage control strategy in distribution networks with high PV penetration. Int J Electr Power Energy Syst 130:106763. https://doi.org/10.1016/j.ijepes.2021.106763

22. Kundur, Prabha S., and Om P. Malik. (2022). Power System Stability and Control. 2nd ed. New York: McGraw Hill.
23. Shuai Z, Sun Y, Shen ZJ et al (2016) Microgrid stability: classification and a review. Renew Sustain Energy Rev 58:167–179. https://doi.org/10.1016/j.rser.2015.12.201
24. Gay P (2018) on good behavior. My Ger Quest 185–206. https://doi.org/10.2307/j.ctt1njk35.14
25. Mumtaz F, Bayram IS (2017) Planning, operation, and protection of microgrids: an overview. Energy Procedia 107:94–100. https://doi.org/10.1016/j.egypro.2016.12.137
26. Dagar A, Gupta P, Niranjan V (2021) Microgrid protection: a comprehensive review. Renew Sustain Energy Rev 149:111401. https://doi.org/10.1016/j.rser.2021.111401
27. Chandak S, Rout PK (2021) The implementation framework of a microgrid: a review. Int J Energy Res 45:3523–3547. https://doi.org/10.1002/er.6064
28. Chandra A, Singh GK, Pant V (2021) Protection of AC microgrid integrated with renewable energy sources—a research review and future trends. Electr Power Syst Res 193:107036. https://doi.org/10.1016/j.epsr.2021.107036
29. Lagos D, Papaspiliotopoulos V, Korres G, Hatziargyriou N (2021) Microgrid protection against internal faults: challenges in islanded and interconnected operation. IEEE Power Energy Mag 19:20–35. https://doi.org/10.1109/MPE.2021.3057950
30. Mohammadi J, Member S, Ajaei FB, Stevens G (2018) Grounding the AC microgrid. 9994:1–9. https://doi.org/10.1109/TIA.2018.2864106
31. Economics and Industry Standing Committee (2019) Implications of a distribution energy future. Western Australia
32. Hashemi SM, Vahidinasab V (2021) Energy management systems for microgrids. In: Anvari-Moghaddam A, Abdi H, Mohammadi-Ivatloo B, Hatziargyriou N (eds) Microgrids: advances in operation, control, and protection. Springer International Publishing, Cham, pp 61–95
33. Wu N, Wang H (2018) Deep learning adaptive dynamic programming for real time energy management and control strategy of micro-grid. J Clean Prod 204:1169–1177. https://doi.org/10.1016/j.jclepro.2018.09.052
34. Ji Y, Wang J, Xu J et al (2019) Real-time energy management of a microgrid using deep reinforcement learning. Energies 12. https://doi.org/10.3390/en12122291
35. Shen J, Jiang C, Liu Y, Qian J (2016) A microgrid energy management system with demand response for providing grid peak shaving. Electr Power Compon Syst 44:843–852

Chapter 7
Standards and Regulations

7.1 General

Standards and guidelines can play a significant role in empowering more transactive and community-oriented energy systems by providing interoperability among technologies, delivering compatible structures to design and implement resilience, secure and safe power supply. The integration of new technologies and DERs, and interoperability appear to be the basis of the performance of future power systems and microgrids as their inevitable constituent. Therefore, standards should be able to guideline the integrating and interaction of DERs and new technologies in the microgrids.

This section aims to introduce the most significant standards around the country and worldwide that might be exerted in microgrids and DERs applications. It is worth mentioning that older standards, most of which are currently withdrawn and inactive, were developed, assuming the comparatively low penetration of DERs in the power system. However, the increasing penetration of DERs raises the need for new standards to preserve the acceptable stability level of a grid. The vast range of standards that cover similar topics may make developers confused as they are slightly inconsistent in some cases.

7.2 Specific Standards for Microgrids

Standards build a base of understanding the requirements of microgrids. Therefore, they are developed to ensure the reliability of supply by ensuring the continuity of supply, power quality, safety and protection, and secure operation. In this subsection, the specific national and international standards for microgrids are covered.

I. A. Ibrahim et al., *Microgrids Design and Operation*, Power Systems,
https://doi.org/10.1007/978-981-96-4914-3_7

7.2.1 *National (Australian) Standards*

The national Australia's standards have developed a fundamental understanding on the requirements of smart grids and microgrids as special kind of smart grids. The microgrid national standards include the requirements of the interoperability and the interconnection of DERs through inverters in grid-connected and islanded microgrids [1, 2].

7.2.1.1 AS/NZS 4777

The AS/NZS 4777 standard [3, 4] focuses on the connection of energy systems via inverters to grids and is comprised of two parts: AS/NZS 4777.1:2016 "Grid connection of energy systems via inverters, Part 1: Installation requirements" [3]; and AS/NZS 4777.2:2020 "Grid connection of energy systems via inverters, Part 2: Inverter requirements" [4].

The requirements of the electrical and general safety installation for inverter energy systems (IES) on LV grids with a rating of up to 200 kVA are mentioned in AS/NZS 4777.1 2016. This standard is utilised along with the technical requirements of the local electricity legislations and appropriate distributors.

The normal performance of LV inverters, commonly used for residential or small-scale commercial grids has been specified in AS/NZS 4777.2, which should be read in adherence with the wiring rules standard, AS/NZS 3000. AS/NZS 4777.2:2015 deserves consideration that the connection needs to be approved by the regulations and installation rules developed by the electricity distributor [5]. The updated version of AS/NZS 4777.2 developed in 2020, provides the capabilities and requirements of the secure operation of a system with high-penetration DERs integration, like distributed solar PV as the most common ones, according to global best practices [6].

7.2.1.2 AS/NZS 4755

The AS/NZS 4755 standard, entitled "demand response capabilities and supporting technologies for electrical products" [7–11], delivers the capabilities of demand response programs (DRPs) as well as the modes of smart devices like DERs, ESSs, air conditioning and pieces of equipment related to electric vehicles. This standard also helps users retain minimum system requirements for demand/generation balancing by using financial incentives payable to consumers to increase, decrease or shift the load demand. In addition, the capability of supporting the security of the power system by using these types of incentives has also been considered through the VPPs concept. On that account, VPPs provide ancillary services, especially frequency control through aggregated distributed resources. It should be taken into account that in case a device is being aimed at contributing to the system security, it must at least have the ability to reduce/shift its power use and start/stop using power during a demand response event. In addition to supporting system security, this capability enables devices to reduce the cost of electricity consumption.

7.2.2 International Standards

International standards refer to the standards that are globally acceptable. They are also more popular among the power developer community. In this regard, the International Electrotechnical Commission (IEC) and IEEE are well known for developing appropriate standards for different aspects of power systems [12].

Regarding best practices, the IEEE 1547 [13] and 2030 [14] standards highlight the interactive interconnection of the grid's components, as well as the need of integrating communications, information technologies (IT) and power under the smart grid concept. Both the IEEE 1547 and IEEE 2030 series of standards, qualified as American national standards, have been sponsored by the Standards Coordinating Committee 21 (SCC21) and approved by the IEEE Standards Board [15, 16]. Accordingly, Table 7.1 lists the IEEE 1547 series of standards.

In addition, IEEE Std. 2030, entitled "Guide for Smart Grid Interoperability of Energy Technology and Information Technology Operation with the Electric Power System (EPS), and End-Use Applications and Loads", has been published in 2011. The IEEE Std. 2030 helps users understand the concept of smart grid interoperability. To accomplish seamless operation, the energy technology, communication, and IT should be properly integrated. Accordingly, Table 7.2 provides different parts of this series of standards.

Furthermore, IEC has developed several standards and series of standards related to planning and operating microgrids as well as integrating DERs with the grid. In this context, the microgrid specific standards by IEC are listed in Table 7.3.

In Europe, CLC EN 50549-1 standard [43], entitled "Requirements for generating plants to be connected in parallel with distribution networks" published in 2019, provides necessary requirements for power plants integration, which are able to work in parallel with the LV grid. This standard or technical reference which is in line with "Commission Regulation (EU) 2016/631: Requirements for Generators", plays a role in agreements between electric power producers and network operators.

7.3 National and International Power Grids Standards

Two major parts of power systems, the transmission and distribution networks are regulated in accordance with their specific standards. As a matter of fact, heavy industrial loads and HV power stations should be operated based on the transmission network standards. However, the regulations set out by state governments are applicable for distribution networks considering different types of consumers. Moreover,

Table 7.1 IEEE 1547 series of standards

Standard	Year	Title	Scope
1547 [13]	2018	"IEEE standard for interconnection and interoperability of distributed energy resources with associated electric power systems interfaces"	Addresses principles for DER interconnection and storage systems. This series of standards also considers the involvement of power inverters/converters and synchronous/induction machines
1547.a [17]	2020	"IEEE standard for interconnection and interoperability of distributed energy resources with associated electric power systems interfaces—amendment 1: to provide more flexibility for adoption of abnormal operating performance category III"	N/A
1547.1 [18]	2020	"IEEE standard conformance test procedures for equipment interconnecting distributed resources with electric power systems and associated interfaces"	Establishes a benchmark for components that might be used in DERs interconnection systems. This standard can be considered as a way of verifying tests by utilities and providers
1547.1a [19]	2015	"IEEE standard conformance test procedures for equipment interconnecting distributed resources with electric power systems—amendment 1"	N/A
1547.2 [20]	2008	"IEEE application guide for IEEE Std. 1547(TM), IEEE standard for interconnecting distributed resources with electric power systems"	Gives information on different DER technologies and the issues caused by their interconnection to the grid. It also helps users understand the background and details of IEEE standards 1547
1547.3 [21]	2007	"IEEE guide for monitoring, information exchange, and control of distributed resources interconnected with electric power systems"	Provides information on different approaches that DER installation stakeholders can adopt to monitor, exchange information, and control their assets. It also helps stakeholders to handle the transactions associated with DERs

(continued)

Table 7.1 (continued)

Standard	Year	Title	Scope
1547.4 [22]	2011	"IEEE guide for design, operation, and integration of distributed resource island systems with electric power systems"	Develops operation and design guidance on intentionally islanded systems known as microgrids and also good practices for DERs integration. There are some criteria that are necessary for an electric system to be included in this guide. First, it must contain DERs and loads as well as a local system. Second, it must be capable of islanding/disconnecting from the grid and working in parallel. Finally, it must be designed deliberately according to the above capabilities
1547.6 [23]	2011	"IEEE recommended practice for interconnecting distributed resources with electric power systems distribution secondary networks"	Presents recommended practices and technical guidance for DERs interconnection to a local electric power system. It also specifies the safety requirements and considerations in accordance with IEEE Std. 1547
1547.7 [24]	2013	"IEEE guide for conducting distribution impact studies for distributed resources interconnection"	Covers the necessary criteria to examine the possible impacts of a single DER or aggregated DER system on the local power system. This guide is beneficial to operators, owners, and interconnection contractors. It provides a detailed methodology that helps them realise when and how they should evaluate the system performance by gathering necessary data and performing required tests
1547.9 [25]	2019	"Guide to using IEEE standard 1547 for interconnection of energy storage distributed energy resources with electric power systems" (draft)	Develops the necessary guidance on the interconnection of energy storage systems as a distributed resource of energy connected to the local energy system. The information provided covers bi-directional power flow (active and reactive) with the ability to export active power, non-exporting storage systems such as UPS and EVs charging stations, and their impacts on the frequency variations

power system security standards, including the standards for power system frequency, and for power system reliability are very important to support/upgrade the existing infrastructure from the security and quality point of view. They also help operators manage the assets in order to connect DERs and grid-side ESSs to either transmission networks or distribution networks. Accordingly, Table 7.4 lists the national and international power grid important categories with their respective standards.

Table 7.2 IEEE 2030 series of standards

Standard	Year	Title	Scope
2030 [14]	2011	"IEEE guide for smart grid interoperability of energy technology and information technology operation with the electric power system (EPS), end-use applications, and loads"	Presents best practices and possible approaches to establish synergy between various components of a smart grid. In fact, it paves the way toward smart grid interoperability by providing a comprehensive framework considering communication, information exchange and control technologies
2030.1 [26]	2015	"Technical specifications of a DC quick and bidirectional charger for use with electric vehicles"	Provides the specifications of the EVs related infrastructures, including requirements for charging stations and supporting pieces of equipment. It also addresses the possible impacts of personal and mass road-based transit on the power system
2030.2 [27]	2015	"IEEE guide for the interoperability of energy storage systems integrated with the electric power infrastructure"	Addresses the necessary set of requirements for hybrid storage systems inserted in the electric power infrastructures considering loads and demand-side applications
2030.2.1 [28]	2019	"IEEE guide for design, operation, and maintenance of battery energy storage systems, both stationary and mobile, and applications integrated with electric power systems"	Provides guidelines for both mobile and stationary battery storage systems and their design, operation, and maintenance. It also addresses the interconnection of these resources to various networks, including distribution and transmission networks, as well as at customer facilities. This guide is beneficial to storage system manufacturers and providers, interconnection contractor, and operators
2030.3 [29]	2016	"IEEE standard test procedures for electric energy storage equipment and systems for electric power systems applications"	Provides methods and strategies for testing of installation and pieces of equipment included in energy storage systems, which can be either a single integrated device or a collection of components that each undertakes a specific task and works together as a system
2030.4 [30]	2017	"Guide for control and automation installations applied to the electric power infrastructure"	Describes the smart grid interoperability reference model (SGIRM) in IEEE Std. 2030 and how to apply it to develop automation components in control systems

(continued)

Table 7.2 (continued)

Standard	Year	Title	Scope
2030.5 [31]	2018	"Standard for smart energy profile protocol"	Focuses on various possible architectures for direct or home area network (HAN) communication between consumers with the ability to export electricity power, also known as prosumers, and a service provider. The aim of this standard is to build a framework for different demand-side energy management strategies and capabilities, including demand response, energy storages, electric vehicles, and utilising alternative carriers such as water and natural gas
2030.6 [32]	2016	"Guide for the benefit evaluation of electric power grid customer demand response"	Establishes an all-inclusive procedure to evaluate the benefits of demand response, including energy saving, financial gains, decarbonisation and emission reduction, and other extended and collective benefits through monitoring its implications for the electricity grid. It also provides indices considering several influencing factors such as the demand response type, user category, and weather conditions
2030.7 [33]	2017	"Standard for the specification of microgrid controllers"	Defines the operation methodology and different aspects of the microgrid energy management system, which plays a vital role in proper system performance. It also addresses the challenges associated with the microgrid controllers such as a required element of any kind of microgrids
2030.8 [34]	2018	"Standard for the testing of microgrid controllers" (draft)	Provides information on how microgrid or distribution network operators can quantify the performance, evaluate various control functions of the microgrid controller and run tests for malfunction
2030.9 [35]	2019	"Recommended practice for the planning and design of the microgrid"	Addresses good practices in the microgrid designing and planning purposes, which applicable to various types of microgrids, including DC, AC, hybrid, intentionally islanded and grid-connected ones
2030.10 [36]	2021	"Standard for DC microgrids for rural and remote electricity access applications"	Establishes the architecture of a DC microgrid that is applicable to rural and remote areas with a nominal distribution voltage of 48 V. It defines voltage and power quality metrics for power supplied to loads attached to such a microgrid. This standard focuses on the power distribution portion of a microgrid and addresses sources only in the way that they are attached to the grid
2030.11 [37]	2021	"IEEE guide for distributed energy resources management systems (DERMS) functional specification"	Describes the grid services provided by DERs which can be integrated into either distribution or transmission networks, with a special focus on DERMS. It also addresses the challenges associated with DERMS interoperability and required infrastructure for communication and data exchange as the ubiquitous feature of modern power grids

Table 7.3 IEC microgrid standards

Standard	Year	Title	Scope
IEC TS 62898-1 [38]	2017	"Guidelines for microgrid projects planning and specification"	Includes guidelines and specifications for planning purposes of LV/MV AC microgrids considering loads and DERs
IEC TS 62898-2 [39]	2018	"Guidelines for operation of microgrids"	Provides guidance for operators to be able to perform the AC microgrid operation
IEC TS 62898-3-1 [40]	2020	"Guidelines for the specification of fault protection and dynamic control in microgrids"	Addresses the protection and dynamic control of grid-connected/islanded AC microgrids with single connection point to the main distribution grid
IEC TS 62786 [41]	2017	"Distributed energy resources connection with the grid"	Specifies the requirements and procedures for connecting DERs to distribution networks, which are applicable to design and operation purposes. This standard covers a wide range of guidelines from device choice, power quality improvement, reactive/active power, and voltage/frequency control, down to the enhancement of monitoring and communication systems and interface protection. The necessary tests that should be run for DER connection are also included
IEC TS 62257 series [42]	2015–2022	"Recommendations for renewable energy and hybrid systems for rural electrification"	This technical specification contains 12 parts with parts 2–6 introducing the management and implementation procedures, while others provide technical specifications for system components working either individually or collectively

Table 7.4 National and international power grids standards

Category	Description	Table
Terminology standards	Define vocabularies related to power grids, especially smart grids	Table C.1
Electrical system operation standards	Cover standards, guidelines, and technical specifications and procedures related to all issues associated with (interconnected) power system operation and control. Security and quality topics are also included in this category	Table C.2
Generation technologies (solar, marin, wind, hydro) standards	Define the standards across various generation technologies and their integration to the grid or user system, such as wind, thermal (including steam and gas/combustion turbines technologies fuelled by oil, gas, and coal), marine (including wave and tidal energy), hydropower, and solar (including rooftop, individual panels and farm installations) energy generation	Table C.3
Asset management standards	Address the good facility management practices to provide the highest financial return, especially via productive assets. The aim of standards in this category is to keep facilities working at their highest capacity in terms of the quality of services provided	Table C.4
System security standards	Define the security issues of not only individuals and staff but also hardware (equipment and network) and software (programs, applications, and data) over various physical conditions and threats (fire, natural disaster, vandalism etc.) in order to avoid considerable losses and damages to entities	Table C.5
Critical infrastructure resilience standards	Include the standards and guides to retain the continuous system function with, at least, the minimum service quality in case of emergency, natural disasters, and other High-Impact-Low-Probability (HILP) events. This category also provides guides to recover the whole system performance (including communications, physical facilities, and supply chain) as quickly as possible after HILP events and conform the system to the new operating conditions, with the aim of preventing a huge impact on the nation in terms of security and social/economic well-being	Table C.6
Substations standards	Cover issues related to substations where the voltage level is changed by transformers. Substations transform the voltage to higher or lower levels to make major facilities with different voltage levels connected to each other. Several substations may be employed to transform the voltage in multiple steps in order for the electricity power to flow throughout the system	Table C.7

(continued)

Table 7.4 (continued)

Category	Description	Table
Transformers standards	Define electrical power transformers as static devices which are able to transform electrical energy by mutual induction among two windings, each connected to a particular circuit. These devices transfer electrical energy from an AC circuit to another without any frequency changes and generally with alteration in the voltage level	Table C.8
Switchgears standards	Covers standards related to switchgears as one of the important pieces of equipment which can highly impact the reliable electricity grid performance. Circuit breakers and switches, fuses, and other protection and isolation devices used to de-energise, disconnect, or isolate equipment are included in the switchgear. This type of equipment plays a role in fault clearance over the downstream network	Table C.9
Protection and relays standards	Address protection relays, a smart device with the ability to make a decision or alert the user by continuously comparing the input and a pre-adjusted set point. Relays can receive voltage, current, temperature etc. as an input and provide visual warning signals as an output for turning power on/off manually or automatically	Table C.10
Cable and overhead lines standards	Define lines as the key parts of a power grid which are responsible for delivering electrical power to consumers and end-users. This category addresses above ground suspended electric lines	Table C.11
Grid side energy storages standards	Define grid-side or in-front-of the-meter energy storage battery systems employed to remove peak loads or compensate for low generation during the system unbalance	Table C.12
Process automation systems standards	Include the standards related to automatic industrial process control carried out by process automation systems (PASs) as an interconnected set of sensors, actuators, and controllers	Table C.13
Electro-magnetic capability standards	Consider the electromagnetic compatibility of devices, as the ability to serve in a specific environment containing electromagnetic radiation while keeping their acceptable and adequate level of performance	Table C.14
Power quality standards	Provide benchmarks for evaluating the power quality attributes, including voltage, current and frequency measured at a specific point	Table C.15

7.4 National and International Supporting Technology Standards

The high expansion of DERs installations, and the development of ESSs, EVSE, new types of generation, smart inverters, advanced controllers, and the emerging technologies (e.g., internet of things (IoT), and digital twin), as well as consumer preferences, are among the major factors and motivations for developing innovative solutions applicable to existing infrastructure. Undoubtedly, sharing the data is a keystone to reinforce the above-mentioned achievements. Here, Table 7.5 gives information on the categories of supporting technologies considering their associated standards. The national and international standards provided support the synergy between existing and developing technologies, as well as enable developers to build up frameworks for design, operation, and implementation to ensure the favourable performance of the whole system.

Summary

Numerous standards have been developed to ensure the safe performance of power systems and their components by IEEE, IEC, and Standard Australia. In general, these standards can also be utilised when it comes to microgrids. However, due to the high penetration of DERs, including new generation technologies and storage systems, there was a need to establish particular standards for microgrids to mitigate DERs' consequential impacts on the system.

Accordingly, several special standards have been developed so far with the IEEE 1547 being the most popular one in the microgrid area according to the survey of national and international projects. The focus of this series of standard is on the interconnection of DERs with the grid, which has some limitations in operating DERs in islanded mode (e.g., IEEE 1547 do not require DER in grid-connected mode to regulate the negative sequence voltage. Instead, it is assumed that the negative sequence voltage is regulated by the substation and the generation resources in the transmission system. However, this is no longer true in a microgrid, where the negative sequence voltage caused by unbalanced load must be controlled by the generation sources inside the microgrid). In addition, IEEE 2030 and IEC 62898 are also among the most common series of standards focusing on microgrids and associated topics. Further, there are numerous requirements and issues related to either power systems in general or microgrid supporting technologies. Regarding national standards, the AS/NZS 4777 and AS/NZS 4755 standards cover some of the functional areas in microgrids. In this context, several effective and applicable standards have been developed by international organisations, national entities follow those standards in case there is no relevant Australian standard. On that account, standards provided by IEC have priority over the standards provided by IEEE. However, it should be noted that the emerging technologies, such as the requirements

Table 7.5 National and international supporting technologies standards

Category	Description	Table
Market systems standards	Include sub-topics on different power and energy market management systems aimed at billing and trading energy, metering, monitoring, and controlling power and energy consumption	Table C.16
Cyber security standards	Provide good practices, technical procedures, and introduction to technological elements used in developing protection systems and supporting cyber security of enterprises, computerised systems and networks, and programs. The protection systems mentioned in this category are responsible for preventing unauthorised access or malicious activity over the dominant system	Table C.17
Local energy storages standards	Address behind-the-meter ESSs and guidelines for their installation at customer premises, connection to the grid, maintenance, and demand management. This category also covers the issues related to recycling battery storage systems	Table C.18
Distributed energy coordination standards	Define the devices used to add information capabilities to various components of distributed energy facilities. Metering, energy-storing, and management devices associated with distributed generation are included in this category	Table C.19
Building management system standards	Refer to all technologies and pieces of equipment used in buildings, plants, and outdoor establishments to develop automated management and control systems	Table C.20
Customer energy management standards	Define customer energy management systems used for monitoring and remote controlling the variables related to energy consumption, such as usage data and energy pricing information. Customer energy management systems are also capable of making decisions manually or automatically to provide an effective consumption control	Table C.21
Demand response management equipment standards	Include an introduction to the instruments used for demand response management systems aimed at demand charge reduction by providing load shifting and peak mitigation services, especially during load peak hours	Table C.22
Advanced metering standards	Refer to a wide range of instruments, elements, software, and communication devices associated with metering infrastructure. This category includes systems such as meter data management (MDM) which are capable of measuring, data gathering and analysis, and generating control signals	Table C.23

(continued)

Table 7.5 (continued)

Category	Description	Table
Electric vehicle supply equipment standards	Address electric powered road vehicles and their charging infrastructure as an emerging part of modern power grids. V2G technology, EV charging facilities, BESS efficiency and recycling and safety issues are among the sup-topics included in this category	Table C.24
Inverters standards	Covers behind-the-meter inverters as a device with the ability to establish a connection between DC and AC circuits by transforming DC to AC	Table C.25
Communication standards	Address communication systems (hardware and software) as one of the most important constituents of power networks and microgrids, which play a significant role in their effective performance. This category embraces several sub-topics such as telecommunication, data channels, public/private communication networks, and associated protocols, to name a few	Table C.26
Data (frameworks and privacy) standards	Refers to the methodologies and protocols of safe data exchange or data flow throughout a power system. User and industry interface, billing and payments, and privacy are addressed in this topic	Table C.27
Digital twin' standards	Define a digital replica of a physical asset/system/process with a bi-directional and real-time communication link. This link helps the former represent the actual behaviour and dynamics of the latter	Table C.28

for grid-forming inverters (e.g., voltage harmonics regulation, frequency, and voltage regulation, maintaining stability, fault ride-through and black start) and hydrogen storage systems are not well covered in the current published standards. This raises the need for a consistent standard framework for microgrid applications, which should be addressed by responsible organisations to either develop new standards or update their existing standards.

References

1. Standards Australia (2016) Standards and the future of distributed electricity, pp 1–36.
2. Standards Australia (2017) Roadmap for standards and the future of distributed electricity
3. Standards Australia (2016) Grid connection of energy systems via inverters—Part 1: Installation requirements. AS/NZS 4777.1:2016
4. Standards Australia (2015) AS/NZS 4777.2:2015: grid connection of energy systems via inverters—Part 2: Inverter requirements

5. Standards Australia. www.standards.org.au
6. AEMO (Australian Energy Market Operator) DER Standards and connections. https://aemo.com.au/en/initiatives/major-programs/nem-distributed-energy-resources-der-program/standards-and-connections
7. Standards Australia (2017) Demand response capabilities and supporting technologies for electrical products. AS/NZS 4755.1:2017 1–36
8. Standards Australia (2016) Demand response capabilities and supporting technologies for electrical products, Part 3.5: Interaction of demand response enabling devices and electrical products—operational instructions and connections for grid-connected electrical energy storage EE. AS/NZS 4755.3.5:2016 1–76
9. Standards Australia (2014) Demand response capabilities and supporting technologies for electrical products, Part 3.1: Interaction of demand response enabling devices and electrical products—operational instructions and connections for air conditioners. AS/NZS 4755.3.1:2014
10. Standards Australia (2014) Demand response capabilities and supporting technologies for electrical products, Part 3.3: Interaction of demand response enabling devices and electrical products—operational instructions and connections for electric storage and electric-boosted storage. AS/NZS 4755.3.3:2014
11. Standards Australia (2014) Demand response capabilities and supporting technologies for electrical products—Part 3.2: Interaction of demand response enabling devices and electrical products—operational instructions and connections for devices controlling swimming pool pump-un. AS/NZS 4755.3.2:2014 1–40
12. IEC. www.iec.ch
13. IEEE (2018) IEEE standard for interconnection and interoperability of distributed energy resources with associated electric power systems interfaces. IEEE Std 1547-2018 (Revision IEEE Std 1547-2003) 1–138
14. IEEE (2011) IEEE guide for smart grid interoperability of energy technology and information technology operation with the electric power system (EPS), end-use applications, and loads. IEEE Std 2030-2011 1–126
15. IEEE Standards Coordinating Committee 21 (SCC21). http://grouper.ieee.org/groups/scc21/index.html
16. Basso T, DeBlasio R (2012) IEEE smart grid series of standards IEEE 2030 (interoperability) and IEEE 1547 (interconnection) status. National Renewable Energy Lab. (NREL), Golden, CO
17. IEEE (2020) IEEE standard for interconnection and interoperability of distributed energy resources with associated electric power systems interfaces—amendment 1: to provide more flexibility for adoption of abnormal operating performance category III. IEEE Std 1547a-2020 (Amendment to IEEE Std 1547-2018) 1–16
18. IEEE (2020) IEEE standard conformance test procedures for equipment interconnecting distributed energy resources with electric power systems and associated interfaces. IEEE Std 1547.1-2020 1–282
19. IEEE (2015) IEEE standard conformance test procedures for equipment interconnecting distributed resources with electric power systems—amendment 1. IEEE Std 1547.1a-2015 (Amendment to IEEE Std 1547.1-2005) 1–27
20. IEEE (2009) IEEE application guide for IEEE Std 1547(TM), IEEE standard for interconnecting distributed resources with electric power systems. IEEE Std 1547.2-2008 1–217
21. IEEE (2007) IEEE guide for monitoring, information exchange, and control of distributed resources interconnected with electric power systems. IEEE Std 1547.3-2007 1–160
22. IEEE (2011) IEEE guide for design, operation, and integration of distributed resource island systems with electric power systems. IEEE Std 1547.4-2011 1–54
23. IEEE (2011) IEEE recommended practice for interconnecting distributed resources with electric power systems distribution secondary networks. IEEE Std 1547.6-2011 1–38
24. IEEE (2014) IEEE guide for conducting distribution impact studies for distributed resource interconnection. IEEE Std 1547.7-2013 1–137

25. IEEE (2021) IEEE draft guide to using IEEE Std 1547(TM) for interconnection of energy storage distributed energy resources with electric power systems. IEEE P1547.9/D5.1 1–80
26. IEEE (2016) IEEE standard technical specifications of a DC quick charger for use with electric vehicles. IEEE Std 2030.1.1-2015 1–97
27. IEEE (2015) IEEE guide for the interoperability of energy storage systems integrated with the electric power infrastructure. IEEE Std 2030.2-2015 1–138
28. IEEE (2019) IEEE guide for design, operation, and maintenance of battery energy storage systems, both stationary and mobile, and applications integrated with electric power systems. IEEE Std 2030.2.1-2019 1–45
29. IEEE (2016) IEEE standard test procedures for electric energy storage equipment and systems for electric power systems applications. IEEE Std 2030.3-2016 1–72
30. IEEE (2011) Guide for control and automation installations applied to the electric power infrastructure. IEEE Std P2030.4
31. IEEE (2018) IEEE standard for smart energy profile application protocol. IEEE Std 2030.5-2018 (Revision IEEE Std 2030.5-2013) 1–361
32. IEEE (2016) IEEE guide for the benefit evaluation of electric power grid customer demand response. IEEE Std 2030.6-2016 1–42
33. IEEE (2018) IEEE standard for the specification of microgrid controllers. IEEE Std 2030.7-2017 1–43
34. (2018) IEEE Draft standard for the testing of microgrid controllers. IEEE P2030.8/D13 1–43
35. IEEE (2019) IEEE recommended practice for the planning and design of the microgrid. IEEE Std 2030.9-2019 1–46
36. IEEE (2021) IEEE standard for DC microgrids for rural and remote electricity access applications. IEEE Std 2030.10-2021 1–47
37. IEEE (2021) IEEE guide for distributed energy resources management systems (DERMS) functional specification. IEEE Std 2030.11 1–61
38. IEC (2017) IEC TS 62898-1: guidelines for microgrid projects planning and specification
39. IEC (2018) Microgrids—Part 2: Guidelines for operation. IEC TS 62898-2 2018 1–38
40. IEC (2020) Microgrids—Part 3-1: Technical requirements—protection and dynamic control. IEC TS 62898-3-12020 1–42
41. IEC (2017) Distributed energy resources connection with the grid. IEC TS 627862017 1–25
42. IEC (2015) Recommendations for renewable energy and hybrid systems for rural electrification—Part 1: General introduction to IEC 62257 series and rural electrification. IEC TS 62257-12015 1–36
43. CLC—European Committee for Electrotechnical Standardization (2019) Requirements for generating plants to be connected in parallel with distribution networks—Part 1: Connection to a LV distribution network—generating plants up to and including Type B. EN 50549-12019

Chapter 8
Digital Twin and Advanced Applications

8.1 Introduction

The grid ecosystem is moving rapidly away from a displacement and replacement paradigm towards ubiquitous connectivity and technology recombination. Various data will be generated and analysed as well as discrete objects, people, and activities will be more connected to the grid ecosystem than ever before [1]. Moreover, the industrial IoT (i.e., industry 4.0) is expected to highly affect the future systems by establishing new technologies, including big data analytic, IoT technologies, AI, machine learning, onboard computation, and intelligent and fast controllers [2]. This along with technology advancement in digitalisation, enables microgrids system operators to deal with microgrids during its entire life cycle, including planning, operation, maintenance and expanding stage using a new concept so-called digital twin. Digital twin can be defined broadly [3], as:

> … digital twin is defined as software-based abstractions of complex physical systems that are connected to the real system via a communication link to continuously exchange data with the real environment and establish a dynamic digital mirror with a constantly running modelling engine ….

In other words, digital twin concept refers to a virtual image or replica that fully describes an actual production or physical process from the micro atomic level to the macro geometrical level that can serve to make safe, reliable and efficient decision [4, 5]. In fact, the eventual purpose of digital twin is to predict physical outcomes of any decisions. Digital twin appears to be a key element of smart manufacturing systems along with three other aspects of autonomy, connectivity and modularity [6]. Another popular application of digital twin is studying the dynamics governing the interdependencies between elements of a complex system like a smart city [7]. The simulation paradigms of such systems have been evolving during two recent decades. In addition, digital twin can refer to as the latest generation of system simulation and representation methods which is dynamic and intelligent by design. This characteristic is known as the key difference of digital twin with other similar

I. A. Ibrahim et al., *Microgrids Design and Operation*, Power Systems,
https://doi.org/10.1007/978-981-96-4914-3_8

methods. One of the eminent features of digital twin concept is the ability to simulate a system in different time scales relying on data collected from expert knowledge and field experiences. When it comes to power systems, one can describe digital twin as the virtual image of the physical object in the electrical power system, which makes the provided data usable [4]. Digital twinning encompasses gathering, processing, and analysing real-time data streaming and provides a great insight into its current and future operating status [3]. The use of digital twin in microgrid planning and implementation will allow the designers to optimise their design and analyse several case studies in a low-cost low-risk environment. As illustrated in Fig. 8.1, the concept of microgrid digital twin is aimed to control and optimise the physical asset using real-time feedback from the twin [8].

The term digital twin may refer to a simple model or a complex digital asset in industries. To make it clearer, Table 8.1 shows four levels of digital twin sophistication based on its features, namely, (i) pre-digital twin; (ii) digital twin; (iii) adaptive digital twin; and (iv) intelligent digital twin. In fact, the maturity of the digital twin differentiate between the levels of sophistication [9, 10].

Several challenges are currently standing in the way of enabling digital twin, especially on a large scale. To address these challenges, which are mainly related to data and syncing the asset and its replica, enabling technologies of digital twin should be identified. In this context, Table 8.2 lists these technologies that can cope with the challenges for wide-speared application of digital twin concept. For instance, IoT technology linking with AI models brings forth full connectivity and high accuracy with no or little missing data, and blockchain technology provides opportunities for tracking and tracing physical and digital assets [11].

Moreover, standardisation of digital twin is another challenge that deserve consideration. Apart from some standards previously developed in related area, the joint technical committee of the International Organisation for Standardisation (ISO) together with the IEC known as ISO/IEC JTC1 as well as IEEE have started some projects to develop standards for digital twinning [12].

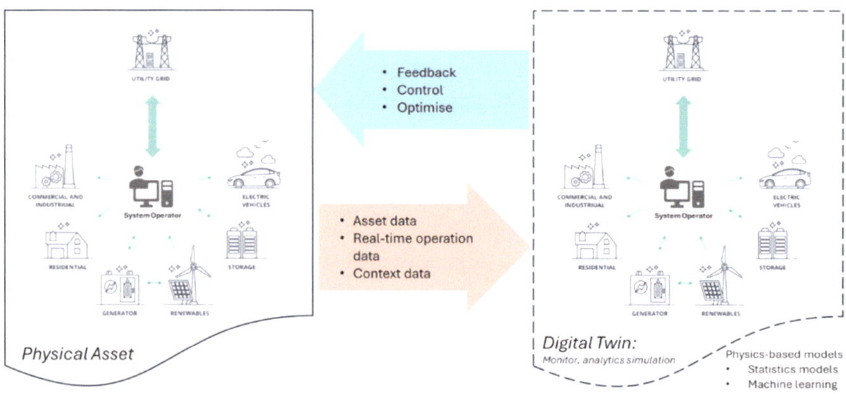

Fig. 8.1 The conceptual microgrid digital twin

Table 8.1 The four levels of maturity of digital twin

Feature	Level 1: Pre-digital twin	Level 2: Digital twin	Level 3: Adaptive digital twin	Level 4: Intelligent digital twin
Physics-based simulation	✓	✓	✓	✓
Physical system	✗	✓	✓	✓
Adaptive graphical user interface (GUI)	✗	✗	✓	✓
Machine learning	✗	✗	✗	✓

Table 8.2 Digital twin concept challenges and enabling technologies

Challenges	Enabling technologies
• Data management • Data privacy • Data security • Data quality	• Digital platforms • Cryptography and blockchain technologies • Big data technologies
• Real-time communication of data • Latency	• Data compression • Communication technologies (e.g., 5G and IoT)
• Physical realism • Future projections	• High fidelity physics-based simulators • Data-driven models • Sensor technologies
• Real time modelling	• Hybrid analysis and modelling • Reduced order modelling • Multivariate data-driven models
• Continuous model updates • Modelling the unknown	• Big data cybernetics • Hybrid analysis and modelling • Data assimilation • Compressed sensing and symbolic regression
• Transparency • Interpretability	• Hybrid analysis and modelling • Explainable artificial intelligence
• Large scale computation	• Computational infrastructure (e.g., Edge, fog, and cloud computing)
• Interaction with physical asset	• Human machine interface • Natural language processing • Visualisation augmented reality • Virtual reality

8.2 Digital Twin Establishment

There are three general parts in digital twinning framework: (i) physical system; (ii) virtual system; and (iii) exchange data between these two systems. Here, Fig. 8.2 shows an overall scheme of digital twin technology with cyber physical security application [10]. As depicted, the digital layer receives data from the physical microgrid and grid communication network.

8.2.1 System Modelling

The following steps are required to model the digital twin environment:

- Real system model. Model the overall system, including all subsystems and their interactions using physics-based/data-driven/hybridisation method which is continuously updated and synchronised [3].
- Physics-based Models. Model the dynamic behaviour of the system using the first principle physical and exact mathematical models to improve the accuracy [13].
- A trade-off between the model accuracy and computation burden needs to be considered as the detailed model can improve the accuracy but will increase its complexity [14], As interoperability of several models and services are important function of digital twin, its fidelity characteristics must be validated before real application.

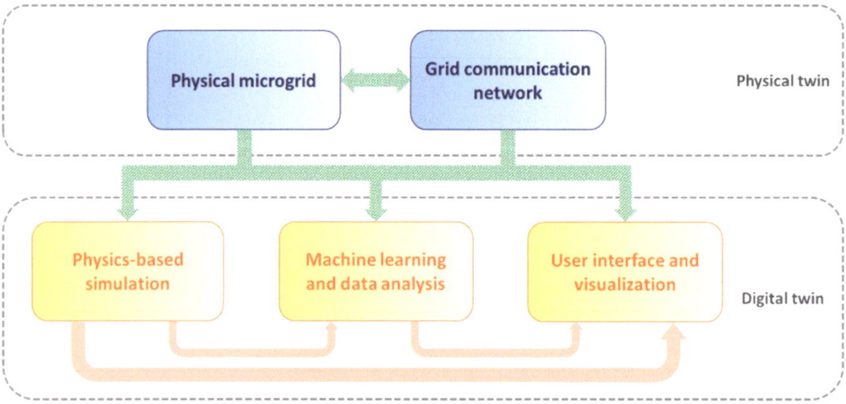

Fig. 8.2 An exemplary schematic of digital and physical twins

8.2.2 Model Adaptation

Model adaptation is utilised to avoid boundary violations considering a certain number of periods. The adaptation process can be realised using the tuned parameters/hyper-parameters of the developed model. There are also other ways to guarantee model adaptation, for example inclusion of more features in data-driven models and integration of optimisation in physics-based models [3]. The model adaptation process can be triggered based on different strategies, meaning that models can be updated with various strategies according to changes in data, such as: (i) fixed periodicity; (ii) event-based approach; and (iii) drastic changes in input data on the basis of defining the trigger [3]. This process should also address the requirements for the services provided by digital twin. A general flowchart of model adaptation is depicted in Fig. 8.3.

8.2.3 Digital-Twining Workflow

The step-by-step workflow of developing a digital twin for microgrids can be briefly listed as follows [3]:

- Digital modelling engine
  - Defining desired services of digital twin and their modelling requirements.
  - Developing models (physics/data-based/hybrid) of the intended assets/subsystems considering the desired level of abstraction.
  - Tuning/validating developed models using historical data collected from physical twin in different operating conditions.
  - Developing models of physical system/process integrating validated models of its subsystems and their interactions.
  - Tuning/validating of the developed system/process models using historical data collected from the physical system in different operating conditions
  - Developing systematic methods for model adaptation based on the twinning error.
- Service development
  - Defining the required algorithms/decision-making process of desired services.
  - Validating/Adjusting of the developed algorithms using simulation environment
  - Defining desired service level (component/system-level), sensitivity to data latency, and required computational platform for deployment of the developed algorithms.
- Service interaction
  - Identifying inter-relation of different services for data exchange.

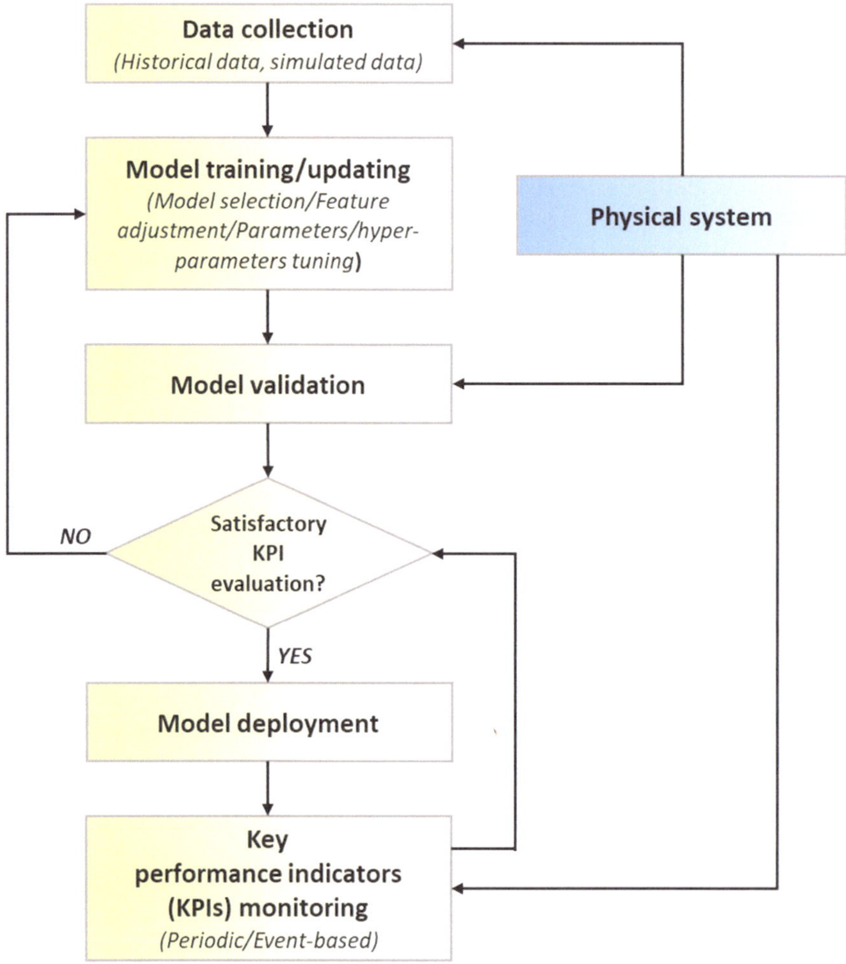

Fig. 8.3 A general flowchart for model adaptation

  - Defining data exchange requirements (type (continues/event-base), rate, etc.).
  - Identifying and deploying required application programming interfaces (APIs) to establish inter-communication of different services.

- Monitoring and communication

  - Design and implementation of the sensor network and data acquisition system for collecting the required data by developed models and methods.
  - Design and implementation of the required communication network for bi-directional data exchange between physical and digital twin.
  - Assessing data quality and implementing required data analysis techniques (outlier detection, de-noising, fusing, visualising, storing, etc.).

- Service assessment
  - Assessing individual service quality with data received from sensor networks and apply required modifications in the sensor and communication systems.
  - Assessing interconnected services performance and making required adjustments.

8.3 Data Management

To synchronise the digital twin with the physical asset to achieve timely and fast detection of events as well as preserving models' accuracy, real-time managing and processing data is substantially important. In this regard, big data analytics provides fast processing and helps extract valuable information during digital twinning [3]. In addition, big data analytics and especially cloud computing can be utilised for several purposes (e.g., data cleaning, reduction of dimension, data fusion, data mining, data organisation, processing, and visualisation). All of these capabilities can boost the real-time data connection [15]. It should be noted that different platforms such as edge, fog, and cloud computing, can be selected based on use case(s). The data includes the historical data obtained from different devices (e.g., smart meter, IoT devices, PMU, etc.) [3].

As mentioned, data is collected from various sources with distinct frequencies. This makes pre-processing, unification, and alignment of the raw data extremely important. Therefore, noisy raw data should be pre-processed by advanced data analysis technologies to provide usable information. Afterward, this information is utilised to update different parts of the system or procedure by sharing with relevant elements [3]. It is worth noting that the intended application of data, including renewable generation monitoring and demand side management (DSM) can differentiate the way of processing [3].

8.4 Communication Protocols

Communication systems are responsible for secure and reliable data transferring between physical microgrids and digital twin. The competence of the communication system are determined based on the quantitative and qualitative requirements of the host system or application, for example reliability, latency, cost, coverage, data rate, interoperability, flexibility, scalability, and security [16]. Considering the purpose of digital twin, various technologies, including wireless technologies (e.g., Wi-Fi, WiMAX, 4G/5G, and satellite technologies) and different wired technologies are available to establish the communication system [16].

8.5 Digital Twin Applications and Use Cases

The digital twin concept is promising research area across power systems and has been infrequently developed in response to the need for facilitating and improving microgrids' design, operation, and control. In recent years some industries, for example General Electric (GE), Siemens, ABB, and Rolls-Royce Marine have adopted digital twin in their operations. GE [17] has developed a digital twin to manage a wind farm, including the topography and environmental information of the wind farms. In 2016, Siemens has developed a digital grid model called "ELVIS" to improve the monitoring and management of Finland transmission system. There are also several applications of digital twin, such as forecasting, planning, asset management, investment plan and operation management of energy supply [18]. Siemens is also among the early adopters of digitalisation and industrial edge technology in drive systems [19]. In marine systems, ABB has developed an innovative digital twin for remote monitoring and predictive maintenance for reliable and economic operations [20]. Moreover, Rolls-Royce Marine has developed an open simulation platform in collaboration with maritime design a digital twin for applications in vessels [21].

In this context, different applications of digital twin are defined based on the model fidelity; therefore, the execution speed based on the application. Accordingly, digital twin can be slower than real-time, execute in real-time, and faster than real-time. If the twin is slower than real-time, it means that the model takes more time to simulate or execute than would pass in reality. For example, to investigate 1 h, it might take 5 h to simulate this one hour. These kinds of twins are usually used for design purposes or to study dynamic phenomena, such as grid stability. However, if the twin runs precisely in real-time, the simulation would be performed in parallel with the reality. This means it can check the system for possible anomalies. In other words, if, for example, the real power plant deviates from what the twin does, there probably exists a problem. If the twin is faster than real-time that can fast-forward into the future to see what will happen or to investigate alternative future scenarios to help in decision making. In this regard, this subsection tries to address and categorise the different potential applications of digital twin in power systems, especially microgrids.

8.5.1 Design and Development

Taking the advantages of digital twinning, the appropriate capacity and size of system components, including generators and DERs, transformers, lines, convertors, ESSs and so forth, can be securely and effectively determined. Moreover, the microgrid digital twin provides the microgrid operators with an opportunity accurate representation of short- and long-term load variations. This capability allows to continuously monitor and reconfigure the system in a low-cost environment with the lowest implementation risk. In general, the microgrid digital twin enables the microgrid operators to run different "what-if" scenarios to check the system behaviour in normal, emergency, and faulty operating conditions before its real implementation. Data obtained from digital twin can be also employed to train machine and deep learning models for different applications as well as to train operators [3].

8.5.2 *Forecasting and Flexibility Identification*

There are several components in a microgrid for example PV systems, wind turbines, BESSs, EVSE, etc. Due to the introduction of smart home appliances and buildings energy management systems, consumers are able to manage their consumption and participate in demand response programs. The intermittent nature of renewable generators and EVs makes forecasting more challenging in microgrids. As a matter of fact, providing more flexibility is the one of the main reasons behind employing these DERs. However, exploring their changing output pattern raises the need for advanced analytics tools. In this context, the identification of the flexibility sources can be supported by the microgrid digital twin in different ways, including:

- Organise and process the historical data and the real-time data stream efficiently while preserving their integrity and privacy.
- Model and simulate the prosumers'/consumers' behaviour in real time.
- Capture the consumers' response to different conditions such as changes in electricity prices, demand response incentives and weather conditions [3].
- Facilitate the systematic integration of different data sources, thus enhancing forecasting accuracy using fusion techniques at the different sensor, feature extraction, and decision levels which improve the forecaster outcome [22].

8.5.3 *Control and Operation Management*

The intelligent control and operation management of digital twin assists operators in several scenarios, for example assessing and managing the mode transitions, steady-state and critical operating conditions, and operating conditions of components, such as their remaining useful life (RUL) and degradation. The performance of different control strategies as well as remote control systems can be evaluated using microgrid digital twin [3]. Providing low latency and massive parallelism, a field programmable gate array (FPGA) can guarantee controllers' real-time operation encountering dynamic changes [23].

8.5.4 Operator Training and Microgrid Autonomy

The microgrid digital twin platform provides an opportunity to upskill the operators of microgrids either humans or machines with a risk-free environment at a reduced cost. More specifically, the microgrid digital twin can be used to broadcast operators' experiences in controlling the system under emergency circumstances as well as to train maintenance services. A human–machine interface (HMI) can be developed using virtual reality (VR) and augmented reality (AR) to facilitate the interaction of the microgrid digital twin and human operators [13]. Thus, microgrid digital twin assist the microgrid autonomous control system to improve its self-healing characteristics by [3]:

- Constituting a true and real-time dynamic behaviour of the physical system and its assets conditions
- Developing advanced simulation platforms to analyse its advancement and predict future evolvement.

Accordingly, the self-healing mechanisms can be activated by a switching control strategy to switch between normal, abnormal and emergency operating strategy [3].

8.5.5 State-of-Health Monitoring and Predictive Maintenance

Digital twinning can be utilised in order to ensure the continued service and optimum conditions for state-of-health (SoH) for infrastructure in microgrids. Given the fact that a huge amount of microgrid costs is devoted to periodic maintenance, it is crucially important to utilise digital twin as a powerful tool for detecting and locating anomalies in asset performance in order to improve efficiency and reduce maintenance cost. The microgrid digital twin provides a platform to closely monitor the state-of-health through HMI and estimate the RUL of components [3].

8.5.6 Fault Diagnosis

As previously mentioned, microgrid digital twin can be utilised to develop the real-time model of a physical system. This capability enables the digital twin to detect any malfunctioning of controllers, sensors and, more importantly, different parts of the main system. When a fault happens in the system, it is vital to follow three consecutive steps as immediately as possible to reduce the system downtime and improve its reliability by identifying faults, fault type, and required actions to mitigate their impacts [3]. Obviously, access to the physical system seems quite arduous in

Table 8.3 The description of security, resiliency, and situational awareness

Term	Description	References
Security	Minimise the impacts of natural disasters and cyber-attacks on critical infrastructures	[24]
Resiliency	Forecast and recover from high-impact low-probability (HILP) with reduced cost and shorter timeframe	[25]
Situational awareness	Identify environmental impacts in short, medium, and longer timeframe	[26]

many instances because of different factors, for example complex installation, high-risk conditions, and costly and laborious procedure. Therefore, microgrid digital twin can be absolutely beneficial by diagnosing faults in a reliable and cost-effective manner as it has highly precision and accurate model of the main system.

8.5.7 Security, Resiliency, and Situational Awareness

As microgrids, similar to any power systems, include two inter-dependent layers, physical and cyber, the associated threats should be addressed in a secure operating plan. To ensure the security of a microgrid, digital twinning could deliver a platform, which timely identifies and simulates possible scenarios of threats, and timely triggers a required action. Data-driven methods or digital twin-based simulation platform can be used to ensure the secure operation of microgrids. Besides, resiliency and situational awareness are among the abilities of microgrids that can be boosted through digital twinning. Accordingly, Table 8.3 provides the definitions of security, resiliency and situational awareness and their differences.

Situational awareness is, in fact, the key aspect of both resiliency and security of the system, which can be supported by digital twin. In this context, digital twin allows a systematically managing enormous data using advanced data analytics methods for pre-processing data, detecting outlier, storing data, etc. Moreover, digital twin-driven simulation platforms along with possible HMI utilisation, delivers accurate current system behaviours and future system trajectories. Further, the digital twin-based decision support system helps microgrid operators to timely react against HILP event propagating quickly [3]. In this context, Table 8.4 summarises the role of digital twin in boosting microgrid resiliency.

8.5.8 Expansion Planning and Policy Making

Power systems are continuously facing changes mainly due to changes in regulations, demands and technologies. This makes the expansion planning extremely attractive to industries. Microgrid digital twin can play a role by modelling and simulating the

Table 8.4 The role of digital twin in improving microgrid resiliency

Phase	Type of action	Key resilience feature	Microgrid digital twin role	Impact
Pre-disturbance	• Preventive actions • Allocating DERs • Re-dispatching Load (considering load priorities) • Reconfiguring system (islanding, clustering, etc.) • Pre-distributing of the assets (crews, mobile generators, mobile ESSs, etc.) • Arranging back up assets	Accurate estimation of event type, location, and its severity	• Deliver a real picture of a current system's state using real-time data analysis, accurate modelling, and estimation techniques • Provide an improved an asset management system using both the historical and real-time data • Enhance the forecasting accuracy of load demand and renewable generation using adaptive data-driven and model-based techniques • Deliver a precise virtual platform to estimate systems' behaviours • Facilitate the interactions among inter-dependent systems and nearby networks	Enhancing operational flexibility

(continued)

Table 8.4 (continued)

Phase	Type of action	Key resilience feature	Microgrid digital twin role	Impact
Disturbance progress	• Emergency actions • DERs re-dispatch Load shedding • Microgrids reconfiguration, islanding schemes, etc.	Advanced monitoring and decision-making systems to notify the evolving situations (sufficient situational awareness)	• Deliver the latest picture of an evolving condition of a system using real-time data connection (as well as soft sensors) and advanced visualisation tools • Enhance integration characteristics of microgrids using timely completion of required evolving tasks	Reducing the speed/level of performance degradation
Post-disturbance	• Emergency coordination • Assessing damage • Identifying available assets for restoration	Disaster assessment and priority setting for quick commencement of system recovery	• Deliver an intelligent decision support system with the capability of prioritising the selection of assets for re-energising and allocation based on offline training • Provide an updated situational awareness and the latest picture of the system states to quantify damage of assets	Fast commencement of system restoration

(continued)

Table 8.4 (continued)

Phase	Type of action	Key resilience feature	Microgrid digital twin role	Impact
Restorative state	• Restorative actions • Re-energising lines • Load restoration • Re-synchronisation	Coordination of restorative actions	• Deliver an intelligent decision support system with the capability of coordinating the restoration actions based on offline training • Evaluate the consequences of actions using an intelligent platform	Fast system recovery
Post-restoration	• Adaptive actions • Evaluation of disaster impacts on systems' performance and resilience • Adaption of required actions for respond for subsequent events	Adaptation of actions during different states	• Self-progressive event model updates the self-regulation of preventive, corrective, and restorative actions to automatically update and re-train the decision support system for up-coming events using real data	Improving system operational and infrastructure resilience

microgrid ecosystem for finding the best time and strategy for the expansion planning. Also, taking the advantages of a testbed provided by microgrid digital twin, the end results of new policies and regulations, and the intention of policymakers are brought as closer as possible. Some examples of applications of the digital twin technologies in the real world are provided in Table 8.5 with the references have been added for further investigation.

8.6 Cybersecurity

Due to digitalisation of microgrids, the cyber security will pose a major threat for its secured and reliable operation. The following four-steps approach can help to establish and maintain cyber-secure systems [39]:

Table 8.5 Examples of digital twin applications in the literature

Area	Application	References
Design, monitoring, and operation management	VPP operation management	[27, 28]
	District EMS	[29]
	Monitor and control of inverters for DERs	[30]
	Monitor and analyse wind farm	[31]
	Observe power transformer	[32]
	Twin renewable generators	[33]
	Twin cyber-physical systems	[34]
	Design and manage microgrids	[35]
	Manage smart home	[36]
SoH monitoring and predictive maintenance	Monitor SoH of Power transformers	[37]
	Detect anomaly of batteries	[38]
	Manage the health of a wind turbine gearbox	[39]
	Estimate RUL of power converters of fixed and floating offshore wind turbines	[40]
	Monitoring SoH of a lithium-ion battery in spacecrafts	[41]
	Monitor battery degradation	[42]
	Manage lithium-ion and lead-acid battery systems	[43]
Fault diagnosis	Diagnose fault of PV systems	[44]
	Detect anomaly of a power converter	[45]
Security, resiliency, and situational awareness	Analyse and mitigate security of microgrids	[10]
	Assess real-time security risk	[46]
	Assess the microgrids controller's performance in improving reliability, resiliency, and efficiency	[47]
	Assess cyber security-false data injection and the denial-of-service cyber-attacks	[48]
	Evaluate microgrids' resilience	[49]

- **Risk assessment**. The first step is to conduct a thorough risk assessment identifying possible internal and external threats.
- **Security policy and processes**. Relevant standards, policies, obligations, and responsibility regarding protection of cyber-physical systems should be clearly communicated to own employees, contractors, and other authorised users.
- **Risk mitigation plan**. Enforce the implementation of an appropriate mitigation plan based on standards (e.g., IEC 62351 and IEEE 1686).
- **Security program**. Manage cybersecurity programs effectively following the previous three points, managing information and communication asset life cycles, understanding of technology upgrade and obsolescence schedules.

8.7 Digital Twin Real-Time Emulation Systems

The real-time emulation systems aim to evaluate the impact of an actual system over a real-time simulated microgrid. This leads to effective planning, operation, and protection of microgrids as well as allows to analyse, predict, and control physical system dynamics.

8.7.1 Real-Time Hardware-in-the-Loop Emulation

Enabled by the hardware-in-the-loop (HIL), it is possible to evaluate the controller performance through various communication and operation scenarios, including large transients and fault conditions, without damaging testing equipment. To capture the accurate dynamic response, sufficient details (e.g., non-linear characteristics of inverters and transformers, switch behaviours, etc.) are required, which are generally difficult to obtain [50]. Several world leading real-time hardware in the loop emulation platforms are used in academic and industrial projects, including Opal-RT, RTDs, Typhoon and RT box.

8.7.2 Hardware Testbed Real-Time Emulation

To conquer the limitations of HIL emulation, a unique test facility called the grid-emulation hardware testbed (HTB) is utilised as another platform for microgrid controller testing. Diverse generation, load, and transmission emulators (e.g., inverters) have been built in the HTB, with a central controller developed in the real commercial NI intelligent electronic devices (IEDs). Unlike HIL, which has a limitation on simulation time scope, the HTB can undertake hours or days long-term real-time emulation while providing detailed live measurement and control data. Using real inverters and switches, the characteristics that cannot be fully captured by the HIL platform can be accurately reproduced [50].

Summary

The advancement of digital technologies as well as the ubiquitous connectivity of smart systems brought up by the industry 4.0 paradigm led to the emergence of digital twin technology. This concept helps make a dynamic digital mirror or a software-based replica of a complex physical system connected to the real system to exchange data through a real-time communication link. The eventual goal of the digital twin concept is to predict and evaluate the outcomes of any decisions. Digital twinning appears to be the latest generation of simulation and representation methods and includes gathering, processing, and analysing real-time data streaming. Obviously, digital twin is beneficial to microgrids for the capability of optimising and analysing the system in a low-cost and low-risk environment. Thus, IEC and IEEE have started developing standards for digital twin technology.

The applications of digital twin in microgrids include, but are not limited to, system design and control, forecasting and flexibility identification, operator training, predictive maintenance, and resiliency improvement. This lead to simplify applying several conditions and tests on the microgrid, which might by risk to apply them in the real microgrids to ensure the reliability and stability of the microgrids. Recently, this promising technology is attracting more attention from the industry side. Based on the reviewed national and international projects, a limited number of projects surveyed have employed digital twin technology, while some of the projects are deploying data acquisition systems with real-time energy management systems platforms, which can be a basis for digital twin implementation in the future. Therefore, linking these platforms with real-time emulation systems through secured communication links can upgrade these platforms to digital twin technology, which can significantly help in improving the operation, reliability, and stability of the microgrids. Also, this can allow the microgrid operator to apply several tests and sever conditions to develop risk management framework.

8.8 Long-Term Planning and Operation Techniques

Machine and deep learning techniques are power tools for long-term planning and operation, which can be used to forecast the future system status, power generation, demand, and energy prices. These aspects can be significantly contributed to optimally operating the microgrid and being ready for any future events as well as they can be helpful for microgrid expansion and upgrades. These techniques can save both time and effort in modelling the components of the microgrids as they do not require explicit models, but their accuracy depends on the volume, resolution and accuracy of the provided data [51]. These techniques are mostly used to forecast the demand/generation profile and energy prices and provide suggestions related to the predictive maintenance of the microgrid components. In addition, several machine learning techniques can be used for clustering and classifying the data, which can lead to optimal data mapping and understanding as well as filling up the missing values and deleting the outlier points. This can lead to more accurate results and reduce the simulation burden significantly. The principle of work for several machine and deep learning algorithms to handle regression, classification and clustering problems are briefly explained in this section as well as their evaluation and validation criteria.

8.8.1 Machine Learning Techniques

Machine learning techniques are one of the applications of AI, where they can use the historical data (i.e., numerical or labels) to characterise the trend of the data into mathematical equations and weights that can be used to predict future values or decisions. Machine learning can be categorised into main types: supervised and unsupervised machine learning techniques. In supervised machine learning techniques (e.g., regression and classification), the inputs and outputs are known; therefore, machine learning techniques used these known variables in training and testing. However, the outputs of the unsupervised machine learning techniques (e.g., clustering) are unknown. Thus, the outputs are grouped in patterns based on a common trend.

8.8.1.1 Regression

Regression is one of the features of supervised machine learning techniques to predict continuous real numbers (e.g., integer and floats) that depend on another variable by iterating over the regression-independent aspect of the data. The regression machine learning techniques can be categorised into several models as follows [52]:

- **Simple linear regression**. The simple liner regression model is used when there is only one continuous variable that is inherently independent to predict the continuous variable that is inherently dependent. Due to the lower computational complexity, linear regression is very fast in predicting the dependent variable compared to nonlinear regression. Additionally, linear regression does not require parameter adjustment.
- **Multiple linear regression**. The multiple linear regression model is used when there are multiple continuous variables that are independent to predict the continuous variables that are dependent. Here, it is necessary to verify that the relationship between each input and each output must be linear before utilising the multiple linear regression model. Verifying the linear relationship between the input(s) and the output(s) can be done through several methods, including the correlation coefficient between output and input. If the correlation coefficient is greater than 0.7, it is safe to assume that there is a linear relationship between the input(s) and the output(s). In addition, if the linear regression model error is unacceptable in any scenario, non-linear regression models must be used to minimise the error. It is necessary to avoid using too many input variables in the multiple linear regression model as this makes the model more sensitive to noise and it also increases the probability of error in the model.
- **Non-linear regression**. Non-linear regression models are mostly applied to predict the continuous dependent variables If the correlation coefficient between input(s) and output(s) is each less than 0.7. The non-linear model can be divided into multiple linear regression models, and then optimise the multiple linear regression models using the least squares optimisation method.

- **Regression trees**. Regression tress models are a type of statistical methods to represent the non-linear datasets with multiple variables and continuous real numbers for prediction purposes by dividing them into multiple samples. The above-mentioned regression models represent the dataset as a whole, but the regression trees models characterise the dataset piecemeal. The regression tree can be used in a case where the modelling of the whole dataset is complex by splitting the entire dataset into multiple samples.
- **Half-sibling regression**. Half-sibling regression models are also a type of statistical methods, which are used to eliminate the systematic noise from observations to reconstruct a latent quantity of interest. Accordingly, the Half-sibling regression models use the information contained in dataset of other observations that are affected by the same noise source to enhance the accuracy.
- **Artificial neural networks (ANNs)**. The generic structure of the ANNs model has input, hidden, and output layers. The ANN structure can have numerous hidden layers. The raw data is received by the input layer and deconstructed by the hidden layers for deep analysis. The goal of the output layer is to identify the result for which the input data is largely similar, and which depends on the pattern of the deconstructed data. ANNs model can also be used in classification.

It is necessary to evaluate the regression model once it has been selected in order to ensure the accuracy of the predict the variable. There are different approaches to evaluating the regression model, including [52]:

- **Mean absolute error (MAE)** is the average absolute error value between the actual measured value(s) and the predicted one(s). The lower MAE means better accuracy.
- **Mean squared error (MSE)** is the mean square of the difference between the measured and predicted value(s), which is an alternative of the MAE. It also has the same reflection on accuracy once MSE is low.
- **Root mean squared error (RMSE)** is introduced as a standard deviation to calculate the errors in the predicted values.
- **Relative absolute error (RAE)** is defined as normalising the residual error summation by dividing it over the total absolute error.
- **Root squared error (RSE)** is the advanced method of evaluating the regression models and is mainly used in big data application.

8.8.1.2 Classification

Classification is another feature of supervised machine learning techniques to categorise the observations into different categories based on training and testing processes. There are numerous classification models used in the literature, including [52]:

- **K-nearest neighbours (k-NN)**. K-NN model identifies an element by finding the value of 'k' that is nearest to the element. The computational burden of this model is low, and it is easy to be implemented.

- **Logistic regression**. The logistic regression model classifies the data points based on the linearity relationships within the dataset. This means that when the datasets are plotted on the x–y axis, there is a clear difference between each sample. The logistic regression model uses the inequalities to calculate the probability of determining the specific category. The decision boundary is the line separating one sample from the other. One of the advantages of the logistic regression model is providing the probability with the categorisation of each new observation.
- **Support vector machine (SVM)**. SVM model can analyse the pattern of data by identifying a decision boundary line between numerous samples in the dataset. SVM model performs very well with a limited amount of data.

A confusion matrix is one of the common assessment methods to validate the accuracy of the classified results. A confusion matrix can be represented by a table, which defines the performance of the classification models.

8.8.1.3 Clustering

Clustering is an advanced feature of the unsupervised machine learning technique that helps to get an intuition about the structure of the data. They divide the unlabelled data into distinct groups based on their unique properties. Therefore, clustering models have several merits that helps to: (i) better analysis and understand the data; (ii) map the data and provide a summary; (iii) remove the outlier points from the datasets; and (iv) identify and remove the repeated data points. These benefits can help in increasing the accuracy of the forecasted results and decreasing the simulation burden [52]. There are two main clustering models are used in the literature, which can be summarised as follows:

- **K-means clustering**. K-means clustering model is mostly applied to large datasets, and it aims to split the untagged data into "k" (i.e., defines the number of pre-defined non-overlapping distinct clusters) matching sub-datasets into several clusters. Therefore, the data within a cluster have similar characteristics, while the data outside the cluster is different. Here, the k-means model is suitable for spherical-shaped clusters or convex clusters, and it can be strongly affected by the presence of noise and outliers in the data.
- **Density-based spatial clustering of application with noise clustering (DBSCAN)**. DBSCAN model can overcome the limitation of the k-means model in handling the nonconvex clusters and outliers/noises. Accordingly, DBSCAN model identifies dense regions, which can be measured by the number of objects close to a given point. This helps in defining the clusters with arbitrary shapes, which k-means model cannot do it.

In general, there is not a solid evaluation method to evaluate the outcome of different clustering models. However, elbow method and silhouette analysis can be used to validate the accuracy of the results and the performance of the models.

8.8.2 Other Machine Learning Techniques

The other machine learning models include the optimised-based machine learning models, where their hyper-parameters are tuned to solve numerous weak points in the present machine learning models for the robustness. The other way to improve the machine learning algorithm is by using ensemble methods. The ensemble methods improve the performance of the machine learning models by combining the forecasting of simple and different models. The quantification method in ensemble methods is done by using a trade-off in bias-variance. A model with high variance improves the model error and a low variance model increases the model error. In addition, bootstrapping methods can be used for the artificial generation of multiple and slightly different datasets for the machine learning models using the bagging concept to improve the accuracy of the forecasted results [52]. The improvements in performance of the machine learning models by one of the above-mentioned methods increase the simulation burden and complexity. Therefore, the selection between a stand-alone machine learning model or a combined model can be done based on the application and a trade-off between accuracy, simplicity, and simulation burden.

8.8.3 Deep Learning Techniques

Deep learning techniques can be considered as advanced machine learning techniques with deeper or higher number of processing layers. The capabilities that make deep learning algorithms and architectures more attractive are unsupervised features learning with big data as well as their generalisation. The most important feature of these techniques is that they do not require any manual feature extractor to deal with regression or classification problems. This concept has been developed to map more complicated non-linear functions that could not be mapped effectively with shallow algorithms [22]. In this subsection, the most popular deep learning techniques are described in brief. According to the literature, all the following techniques are applicable for both of regression and classification problems. In addition, the performance and accuracy of deep learning techniques can be evaluated using the same evaluation criteria that used to evaluate the regression and classification models in machine learning techniques.

8.8.3.1 Convolutional Neural Networks

Convolutional neural networks (CNNs) are widely used in object detection problems through images as well as in time-series forecasting and transient stability analysis [53]. In general, the architecture of CNNs depends on the application (i.e., classification and regression), and can vary based on the number and type of the layers considered [54]. This technique is comprised of multi layers, including normalising and fully connected layers preceded by a series of sub-sampling (i.e., pooling operation) and convolution layers, which are responsible for extracting features from data. The convolution layer in CNNs performs the convolution operation. The other layer is called rectified linear unit, which produces the rectified feature map by performing operation on elements. The output from the rectified linear unit is fed into the pooling layer for dimension reduction resulting in two-dimensional arrays. Next, the pooled feature map is flattened and converted to a continuous linear vector. Finally, the fully connected layer predicts or classifies the objects [55].

8.8.3.2 Recurrent Neural Networks

Recurrent neural networks (RNNs) consist of several connections between nodes to form a directed graph with a feedback loop [56]. The RNN technique is a combination of several feed-forward neural networks. Thus, the output of a layer is fed into the next layer as an input, forming directed cycles to memorise the states and store the past information. RNN is able to generate a sequence of outputs by taking a sequence of inputs at any length. This feature makes it very applicable to problems that need processing sequences of time-phased input data such as time-series analysis [57]. In addition, the model size does not influence the input size because the historical information is considered for computation. However, when a model consists of longer-term dependencies RNNs, it may encounter a problem of very small or large network weights. This problem brings down the efficiency of the learning process. To solve this issue, a novel and commonly applied type of RNNs called long short-term memory (LSTM) is developed which is able to learn long-term relationships with its extra gates [54].

8.8.3.3 Long Short-Term Memory

LSTM [58], is a special supervised algorithm of RNNs technique, that can extract temporal features while learning and memorising long-term dependencies. Therefore, LSTMs are widely used to deal with time-dependent data and classify sequential data [26]. The architecture of LSTM technique is comprised of three gates, each performing a special function. It is initially launched by receiving the information from the previous timestamp, which is decided to be remembered or removed if irrelevant (i.e., forget gate). Afterwards, in the Input gate, new information coming as the input, which should be learned by the current cell. In the third gate (i.e., output gate), the information is updated and passed to the next timestamp [59]. Retaining the knowledge over time makes LSTMs applicable to time-series prediction and other problems that need state awareness [57]. It should be noted that the LSTM approach merely considers one-way propagation.

8.8.3.4 Bi-directional Long Short-Term Memory

Bi-directional long short-term memory (Bi-LSTM) is aimed at two-way input flow, which is a developed version of LSTM. Using this feature, the Bi-LSTM is able to maintain not only the past but also the future information. This algorithm shows a better prediction and generalisation performance than the ordinary one and can deal with more complex models [60]. Regarding the architecture of Bi-LSTM, the 2 LSTM units are merged together, one of which is responsible for backward direction (i.e., future to past), while the other one accomplishing the forward direction (i.e., past to future). The combination of these two networks' output form a new vector and sequentially generate the eventual output [61]. Accordingly, the accuracy of Bi-LSTM is expected to be better than LSTM, while its training time will be longer.

8.8.3.5 Gated Recurrent Units

Gated recurrent units (GRUs) are simplified versions of the LSTMs structure [25]. In fact, GRUs only consider the forget and Input gates and, unlike LSTMs, do not include output gates, resulting in their smaller size than LSTMs. This algorithm also modulates the flow of information without any separate cells for memorising them or any mechanisms that control the degree of state exposure [25].

8.8.3.6 Generative Adversarial Networks

Generative adversarial networks (GANs) are comprised of generative and discriminative neural networks. The former produces new fake information as much deceptive as possible and the latter is responsible for distinguishing between the real input data and fake generated data. Accordingly, the two networks work against each other and that is why this technique is known as "adversarial". The deceiving process will continue until the generator is able to generate synthetic data that is not determined as fake data by the discriminator. In other word, when this process is no longer continued, new synthetic data can be created by the generator. GANs have a variety of applications, including positive ones (e.g., image-to-image, image-to-text translation, and model-free scenario generation) as well as negative ones (e.g., fraud). The larger sharpness of GANs compared to their counterparts is among the merits of this architecture especially for probabilistic applications. However, uncertain convergence and bounded feature diversity can be the main demerits of this technique [26].

Summary

Machine and deep learning techniques are widely used in academic research in several applications, including forecasting numeric values, classifications, and clustering the datasets. Several machine and deep learning models are used to model the microgrid components (e.g., PV systems, wind turbines, etc.) based on historical data. The accuracy of such models strongly depends on the volume, resolution, and accuracy of the training data. However, such models are limited to the characteristic of the component that its data is used in training. Therefore, such a model cannot be generalised for all the microgrids, except the same technology characteristic is deployed. In addition, clustering models are significantly useful in mapping and cleaning the data, which can improve the performance and accuracy of the simulation models in microgrid design processes. However, machine and deep learning techniques are promising models for long-term planning and operation of microgrids using an online learning process based on real-time measures. Therefore, such techniques can significantly support the optimisation of microgrid real-time operation by forecasting the demand/generation profiles and market prices. The online learning process will keep updating the accuracy of such models by applying the feedback correction factors. Therefore, the microgrid control centre and the operator can be ready for any future events in the short-time horizon as well as they can provide suggestions for future microgrid expansion and upgrades. Machine and deep learning techniques are rarely used in national and international microgrid projects. Therefore, more investigations are required to assess the capability of machine and deep learning techniques in several applications in microgrids.

References

1. Schneider Electric (2018) Digital grid unleashed. Andover, MA, USA
2. Goossens P (2017) Industry 4.0 and the power of the digital twin. Retrieved 5
3. Bazmohammadi N, Madary A, Vasquez JC et al (2021) Microgrid digital twins: concepts, applications, and future trends. IEEE Access 10:2284–2302
4. Brosinsky C, Song X, Westermann D (2019) Digital twin-concept of a continuously adaptive power system mirror. In: International ETG-congress 2019; ETG symposium. VDE, pp 1–6
5. Grieves M, Vickers J (2017) Digital twin: mitigating unpredictable, undesirable emergent behavior in complex systems BT—transdisciplinary perspectives on complex systems: new findings and approaches. In: Kahlen F-J, Flumerfelt S, Alves A (eds) Springer International Publishing, Cham, pp 85–113
6. Rosen R, Von Wichert G, Lo G, Bettenhausen KD (2015) About the importance of autonomy and digital twins for the future of manufacturing. Ifac-papersonline 48:567–572
7. Mohammadi N, Taylor JE (2017) Smart city digital twins. Proceedings. In: 2017 IEEE symposium series on computational intelligence (SSCI)
8. Park H-A, Byeon G, Son W et al (2020) Digital twin for operation of microgrid: optimal scheduling in virtual space of digital twin. Energies 13:1–15. https://doi.org/10.3390/en13205504
9. Madni AM, Madni CC, Lucero SD (2019) Leveraging digital twin technology in model-based systems engineering. Systems 7:7

10. Danilczyk W, Sun Y, He H (2019) Angel: an intelligent digital twin framework for microgrid security. In: 2019 North American power symposium (NAPS). IEEE, Wichita, KS, USA, pp 1–6
11. Fuller A, Fan Z, Day C, Barlow C (2020) Digital twin: enabling technologies, challenges and open research. IEEE Access 8:108952–108971. https://doi.org/10.1109/ACCESS.2020.2998358
12. Malakuti S, Schalkwyk P Van, Boss B et al (2020) Digital twins for industrial applications. Definition, business values, design aspects, standards and use cases
13. Rasheed A, San O, Kvamsdal T (2020) Digital twin: values, challenges and enablers from a modeling perspective. IEEE Access 8:21980–22012
14. Guerrero JM, Vasquez JC, Matas J et al (2010) Hierarchical control of droop-controlled AC and DC microgrids—a general approach toward standardization. IEEE Trans Ind Electron 58:158–172
15. Tu C, He X, Shuai Z, Jiang F (2017) Big data issues in smart grid–a review. Renew Sustain Energy Rev 79:1099–1107
16. Ancillotti E, Bruno R, Conti M (2013) The role of communication systems in smart grids: architectures, technical solutions and research challenges. Comput Commun 36:1665–1697
17. Lund AM, Mochel K, Lin J-W et al (2018) Digital twin interface for operating wind farms
18. Siemens Electrical Digital Twin. https://siemens.com/electrical-digital-twin. Accessed 29 Dec 2021
19. Digitalization in Drive Technology. https://new.siemens.com/global/en/products/%0Adrives/digitalization-in-drive-technology/data-analytics.html. Accessed 29 Dec 2021
20. ABB Ability Collaborative Operations for Marine. https://new.abb.com/marine/systems-and-solutions/digital/collaborativeoperations. Accessed 29 Dec 2021
21. Launch of Open Simulation Platform for Creating New Ships. https://www.rolls-royce.com/media/press-releases/2018/21-03-2018-launch-of-open-simulation-platform-for-creating-new-ships.aspx. Accessed 29 Dec 2021
22. Lau BPL, Marakkalage SH, Zhou Y et al (2019) A survey of data fusion in smart city applications. Inf Fusion 52:357–374
23. Milton M, De La OC, Ginn HL, Benigni A (2020) Controller-embeddable probabilistic real-time digital twins for power electronic converter diagnostics. IEEE Trans Power Electron 35:9850–9864
24. Biringer B, Vugrin E, Warren D (2013) Critical infrastructure system security and resiliency. CRC Press
25. Panteli M, Trakas DN, Mancarella P, Hatziargyriou ND (2017) Power systems resilience assessment: hardening and smart operational enhancement strategies. Proc IEEE 105:1202–1213
26. He X, Ai Q, Qiu RC, Zhang D (2019) Preliminary exploration on digital twin for power systems: challenges, framework, and applications. arXiv Preprint arXiv:190906977
27. Tugarinov P, Truckenmüller F, Nold B (2019) Virtual power plant demonstration platform. In: Forum of mining engineers, Dniepro
28. Tugarinov P, Truckenmüller F, Nold B (2019) Digital twin of distributed energy devices. In: Proceedings of the international conference on science and technology, forum mining engineering, pp 323–331
29. O'Dwyer E, Pan I, Charlesworth R et al (2020) Integration of an energy management tool and digital twin for coordination and control of multi-vector smart energy systems. Sustain Cities Soc 62:102412
30. Senthilnathan K, Annapoorani I (2019) Multi-port current source inverter for smart microgrid applications: a cyber physical paradigm. Electronics 8:1
31. Pargmann H, Euhausen D, Faber R (2018) Intelligent big data processing for wind farm monitoring and analysis based on cloud-technologies and digital twins: a quantitative approach. In: 2018 IEEE 3rd international conference on cloud computing and big data analysis (ICCCBDA). IEEE, pp 233–237
32. Moutis P, Alizadeh-Mousavi O (2020) Digital twin of distribution power transformer for real-time monitoring of medium voltage from low voltage measurements. IEEE Trans Power Deliv 36:1952–1963

33. Ebrahimi A (2019) Challenges of developing a digital twin model of renewable energy generators. In: 2019 IEEE 28th international symposium on industrial electronics (ISIE). IEEE, pp 1059–1066
34. Saad A, Faddel S, Mohammed O (2020) IoT-based digital twin for energy cyber-physical systems: design and implementation. Energies 13:1–19. https://doi.org/10.3390/en13184762
35. Béguery P, Pflaum P, Mugnier C (2019) Microgrid energy management optimization—a common platform for research development and design tools. In: Proceedings of the 16th IBPSA conference, Rome, pp 3163–3170
36. Shvedenko VN, Mozokhin AE (2019) Methodological foundations for the formation of information space and digital twin objects in smart homes. Autom Doc Math Linguist 53:303–308
37. Yang Y, Chen Z, Yan J et al (2019) State evaluation of power transformer based on digital twin. In: 2019 IEEE international conference on service operations and logistics, and informatics (SOLI). IEEE, pp 230–235
38. Pileggi P, Verriet J, Broekhuijsen J et al (2019) A digital twin for cyber-physical energy systems. In: 7th workshop modeling simulation cyber-physical energy systems (MSCPES), pp 1–6
39. Tao F, Zhang M, Liu Y, Nee AYC (2018) Digital twin driven prognostics and health management for complex equipment. Cirp Ann 67:169–172
40. Sivalingam K, Sepulveda M, Spring M, Davies P (2018) A review and methodology development for remaining useful life prediction of offshore fixed and floating wind turbine power converter with digital twin technology perspective. In: 2018 2nd International conference on green energy and applications (ICGEA). IEEE, pp 197–204
41. Peng Y, Zhang X, Song Y, Liu D (2019) A low cost flexible digital twin platform for spacecraft lithium-ion battery pack degradation assessment. In: 2019 IEEE international instrumentation and measurement technology conference (I2MTC). IEEE, pp 1–6
42. Baumann M, Rohr S, Lienkamp M (2018) Cloud-connected battery management for decision making on second-life of electric vehicle batteries. In: 2018 Thirteenth international conference on ecological vehicles and renewable energies (EVER). IEEE, pp 1–6
43. Li W, Rentemeister M, Badeda J et al (2020) Digital twin for battery systems: cloud battery management system with online state-of-charge and state-of-health estimation. J Energy Storage 30:101557
44. Jain P, Poon J, Singh JP et al (2019) A digital twin approach for fault diagnosis in distributed photovoltaic systems. IEEE Trans Power Electron 35:940–956
45. Zhou M, Yan J, Zhou X (2020) Real-time online analysis of power grid. CSEE J Power Energy Syst 6:236–238
46. Nowocin JK (2017) Microgrid risk reduction for design and validation testing using controller hardware in the loop. Massachusetts Institute of Technology
47. Saad A, Faddel S, Youssef T, Mohammed OA (2020) On the implementation of IoT-based digital twin for networked microgrids resiliency against cyber attacks. IEEE Trans Smart Grid 11:5138–5150
48. Judah G (2018) Power systems modelling of a community energy project. In: CIRED 2018 Ljubljana workshop. AIM, Ljubljana, Slovenia
49. Godemel F Focus on your risk strategy: cyber security for the smart grid. https://blog.se.com/smart-grid/2017/07/28/focus-risk-strategy-cyber-security-smart-grid/. Accessed 28 Jul 2017
50. EPRI (2018) A smart and flexible microgrid with dynamic boundary and intelligent open-source controller. Palo Alto, California, USA
51. Rouzbahani HM, Karimipour H, Lei L (2020) An ensemble deep convolutional neural network model for electricity theft detection in smart grids. In: 2020 IEEE international conference on systems, man, and cybernetics (SMC). IEEE, pp 3637–3642
52. Rapaport S, Green M (2021) The use of advanced algorithms in PV failure monitoring. Paris, France
53. Shi Z, Yao W, Zeng L et al (2020) Convolutional neural network-based power system transient stability assessment and instability mode prediction. Appl Energy 263:1–13
54. Deep Leaning Toolbox. https://mathworks.com/help/deeplearning/. Accessed 6 Jun 2022

55. Standards Australia (2017) Demand response capabilities and supporting technologies for electrical products. AS/NZS 4755.1:2017 1–36
56. Wang H, Lei Z, Zhang X et al (2019) A review of deep learning for renewable energy forecasting. Energy Convers Manag 198:1–16. https://doi.org/10.1016/j.enconman.2019.111799
57. Shrestha A, Mahmood A (2019) Review of deep learning algorithms and architectures. IEEE Access 7:53040–53065. https://doi.org/10.1109/ACCESS.2019.2912200
58. Chung J, Gulcehre C, Cho K, Bengio Y (2014) Empirical evaluation of gated recurrent neural networks on sequence modeling. arXiv Preprint https://doi.org/10.48550/arXiv.1412.3555
59. Khodayar M, Liu G, Wang J, Khodayar ME (2021) Deep learning in power systems research: a review. CSEE J Power Energy Syst 7:209–220. https://doi.org/10.17775/CSEEJPES.2020.02700
60. Yang YL, Wan GC, Tong MS (2022) A novel wireless propagation model based on Bi-LSTM algorithm. IEEE Access 10:43837–43847. https://doi.org/10.1109/ACCESS.2022.3169174
61. Xueqing Z, Zhansong Z, Chaomo Z (2021) Bi-LSTM deep neural network reservoir classification model based on the innovative input of logging curve response sequences. IEEE Access 9:19902–19915. https://doi.org/10.1109/ACCESS.2021.3053289

Chapter 9
Best Practices in Microgrid Development and Future Research Directions

9.1 Lessons Learnt

Several lessons learnt have been collected and identified from the reviewed national and international projects in Appendices A and B, respectively, to guide the decision making within communities considering microgrid projects. The collected lessons learnt are categorised into five main groups to consider the general aspects as well as design and operation, economic aspects, social and environment lessons learnt.

9.1.1 General Lessons Learnt

The general collected lessons learnt based on the national and international practices can be listed as follows:

- Microgrids have significant benefits in developments of land and housing regarding environment, society, and economy; and considerably enhance the delivery of the national energy objective.
- Microgrids should be implemented in a way that incorporates the goals of developers, wider communities, owners and stakeholders, operators, and residents. Financial returns along with a social license for each stakeholder need to be ensured.
- An improved hybrid multi-user local ownership model of microgrids is essential for its wider adoption in regional areas.
- Further challenges, for example uncertainties in regulation, technology prices, community trust and involvement, and carbon pricing need to be addressed in order to make community microgrids a viable and preferred future energy solution.
- The main drivers for regional community microgrids are security of emergency services, resilience, and sustainability of long-term power supply.

I. A. Ibrahim et al., *Microgrids Design and Operation*, Power Systems,
https://doi.org/10.1007/978-981-96-4914-3_9

- Detailed and thorough cost-benefit analyses are important to optimise microgrid design and operation to maximise its benefits.
- Commissioning, system tuning and testing, as well as adjusting the performance over time should be considered as a whole for the optimised operation to be guaranteed. This deserves even more consideration to validate islanding operation mode of microgrids.
- New technologies necessitate larger-than-usual finance reserves, however, the amount of money provided by grants is fixed. As a result, the project owner should set aside a contingency fund to cover higher project costs resulting from unforeseeable events such as equipment or vendor failure and delays.
- Although incentive-based schemes, along with tariff adjustments, could extend the use of behind-the-meter solar and battery installations, this could lead to some challenges such as significant technical issues and upsetting the balance between the cost and revenue of utilities.
- A reliable and secure communication requirement should be met over the project's lifetime by installing fibre to critical points.

9.1.2 Design and Operation Lessons Learnt

The design and operation collected lessons learnt from the national and international projects are categorised into four main subsections to cover the technical, regulation, safety, and management aspects.

9.1.2.1 Technical Aspects

The technical collected lessons learnt based on the national and international practices can be listed as follows:

- Integrating mix of renewable technologies can enhance the reliability and resilience of a microgrid.
- Flexibility and energy efficiency can play an important role in the transition of renewable microgrids.
- For the secure and stable operation of largely renewable islanded and remote community microgrids, installation of energy storage systems is extremely important.
- Microgrids and batteries assist network operators in managing technical issues associated with intermittency, voltage variability, voltage rise, and unscheduled peak events.
- Data access is a challenging task to model the feasibility study as it requires an access to participants' energy usage data.
- Because of few precedents and many variables, modelling these projects is highly complex.
- Making the data available through open protocols is a significance important.

- In relation to distribution transformer ratings, there may be a greater degree of sensitivity, especially if large scale solar is connected for export.
- Installation and shared energy supply from a hybrid energy system with PV and BESSs can leverage opportunities and mitigate the technical issues in grids.
- Large amounts of renewable energy are challenging to integrate into such power systems without compromising supply or dependability. Overcoming these integration issues will reduce the community's reliance on diesel, lower energy production costs, and lower carbon dioxide emissions.
- ESSs for a microgrid can be installed in a modular way that allows them to be redeployed if they are not needed in the future.
- Providing good quality, trusted online platform of information may be most effective when promoting solar PV systems, BESSs, and other energy products.
- Consider sites when subdivisions are being planned, allowing developers to incorporate the community ESSs into their planning.
- Analysing interval data and patterns of consumption can be used to identify which customers would be the most likely to benefit from a future behind-the-meter battery solution.
- An inverter tripping event has been noticed in a project which could be caused by a sudden load decrease on a particular feeder and its wider post transient effects when high PV generation is available. However, reliable, and robust determination of cause of inverter tripping events is a challenging issue due to limited knowledge to inverter settings and access in controller tunings.
- The frequency of occurrence of high voltage events has a strong and direct correlation with the profile of PV output power. Under-voltage issues could be experienced during peak hours.
- A selective and individualised solution to address voltage issues is required based on individual LV network behaviours. Further, balancing between DER generation and demand across phases should be reviewed. The connection of inadequate conductors between the supply point and switchboard and/or between the switchboard and the inverter-based DER can contribute to worsening the voltage-rise issue. Better strategies need to be applied to deal with this issue.
- The number of required BESSs would be doubled in the absence of the cloud detection system. Moreover, the batteries would need to work at higher discharge rates in order for the output to be maintained within the acceptable ramp rate.
- Considering both generation and demand sides opportunities in the grid management, as well as high-speed communications, provides benefits, including reduced carbon emissions and reduced costs, while improving reliability in the supply side.
- The DER management system can interact with BESSs at customer premises by sending and receiving commands.
- Novel technologies such as ESSs and other enablers, along with containerising power systems can be helpful in increasing the financial efficacy of islanded systems located in a remote area.
- In a real-time digital simulation environment, HIL testing is advantageous. Thus, operators will be no longer exposed to risk and have access to a training facility.

- A smooth transition from the grid-connected to islanded mode necessitates the use of the appropriate equipment, specialised expertise, and meticulous execution.
- The islanding operation in a low inertia environment require changes to protection setpoints and ride-through settings.
- A special interface application is required for all controllers and devices. This may impose extra cost and time to both developing and commissioning processes. However, the standardisation approach will solve this problem.
- The greater automation can considerably increase the system complexity. As a matter of fact, increasing the time interval between grid loss and grid forming from three seconds to three cycles raises commissioning and programming times, as well as project costs and risks.
- Peak demand drops and flattening of the daily demand profile are crucial to the success of the hybrid power plant. Demand-side management strategies, such as power decentralisation and storage, can help to achieve this.
- Having a hysteresis band reduces the frequency with which the battery can be employed when its SOC is near the minimum and maximum limit.
- Due to its ability to store energy for longer periods of time, hybrid energy systems with hydrogen could improve the performance of microgrids.
- To enable the microgrid to provide grid services, the ratio of dispatchable resources like batteries to base load should be greater than one. Furthermore, the system should meet the minimum sizing requirement of the utility.
- Companies with more experience in energy management have more solutions to utilise newer technologies in the best possible way. Moreover, the cooperation of companies creates a greater understanding of the capacities, communication protocols and how their counterparts' equipment works.

9.1.2.2 Regulation Aspects

The collected lessons learnt considering the regulations and standards according to national and international practices are listed as follows:

- Considering industry standard practices, containerising electrical equipment has significant benefits, especially when it comes to factory acceptance testing of complete equipment.
- It is not impossible for distribution network service providers to include BESSs in their regulatory expenditure submissions in the existing regulatory framework.
- Even though service standard targets are meant to provide incentives for distribution network providers to meet and surpass reliability standards, it is difficult to justify expenditures on reliability outliers when service standard benchmarks are being met.
- Since the community ESSs construction is a unique concept, early engagement with local government agencies and regulatory bodies is essential.
- Different regions have different planning processes and approvals, and parties are unlikely to have dealt with this type of proposal before. Neither new types of proposals, such as community ESSs, nor policy frameworks guiding decision

making have been considered. Therefore, early consultation with stakeholders is crucial.

- Inform authorities and stakeholders regularly and, where possible, give visual aids so that roadblocks and gaps in information can be identified early.
- There are substantial differences between performance-based solutions and international standards.

9.1.2.3 Safety Aspects

The safety aspects for humans, animals and devices are very important in implementing, operating, providing maintenance and replacement of the components in the microgrids. Accordingly, the collected lessons learnt related to the safety aspects based on the national and international practices can be summarised as follows:

- Consider all potential risks during site selection, including:
  - Safety of persons
  - Safety for residents
  - Security of the community ESSs
  - Distance from community infrastructures such as schools and parks
  - Access to adequate energy infrastructure
  - Building size and requirements
  - Regulation compliance and approval
  - Ownership and availability of land on appropriate commercial terms
  - Residential housing nearby
  - Site assets
- Sentiment and reactions of the community.
- Chillers and dew point monitors are critical process and safety needs.
- Due to the installation ambiance as well as outdoor temperature of such areas, it is of a lot of importance to provide employees and staff with comfort in order to reduce possible injuries and heatstroke.

9.1.2.4 Management Aspects

The collected lessons learnt considering the management aspects according to national and international practices are listed as follows:

- Extensive first-time mobilisation requirements (e.g., medicals, internal permissions, mine approvals, travel, and lodging bookings) combined with high labour turnover owing to the nature of the work (e.g., repetitive chores in the heat of summer) resulted in a substantial quantity of administrative tasks.
- The microgrid project coordinator must require all vendors to discuss the interaction of their systems with other systems to ensure the best performance and connectivity.

- In projects that require design and coordination at the managerial level and between vendors, a coordinator with comprehensive awareness of the project's components and commissioning plans, must be on the team.
- The operation of a working campus or commercial facility will be impacted by establishment a new microgrid. This necessitates the involvement of the IT and operations teams with control strategies, data network structure, and connection points where appropriate. It is also important to provide the top manager with a genuine picture of concerns and issues that a microgrid project is facing.

9.1.3 Economic Aspects Lessons Learnt

The economic collected lessons learnt from the national and international practices are listed as follows:

- The economy is the primary driver in most of the projects. The cost of producing electricity far exceeds the revenue.
- Retail plans can be confusing.
- Every market has local factors, but its drivers differ, and community can be defined in more flexible manner than only geographic consideration.
- Energy bills should be more informative.
- Many projects are stymied by the cost and regulations of transferring energy across the grid.
- Considering the high level of uncertainty, treat the cost of network connections for new technologies as either provisional or apply a separate contingency.
- Commercial arrangements, especially for embedded generators, should consider all possible events related to placing or lifting constraints and their associated outcome or simply committing to negotiation in those circumstances. This can enable the system operator to control or constrain the embedded generation, as per other technical requirements included in a power purchase agreement and connection agreement.
- There are a variety of business structure options and choosing the right one is critical to minimising taxes and maximising low-cost financing opportunities.
- The selling price of the electricity generated by the hybrid power plant is a major indicator of the project's economic viability.

9.1.4 Social Lessons Learnt

The collected lessons learnt considering the main actors in the microgrid and social aspects according to national and international practices are listed as follows:

- Increased power prices are major concerns for communities.
- Power price changes worry people. In deciding whether to install their own solar system and storage, many people are influenced by uncertainty.

- Power prices are in the spotlight as people worry about how they will change. For many people, uncertainty is a major driving factor in their decision whether or not to install a solar system and storage.
- People are not always rational on financial issues.
- People don't necessarily look for the cheapest energy deals.
- Some people do not understand the local energy marketplace. Therefore, any project should include a program describing clearly what participation entails.
- In order to maximise the benefits of battery storage, participants must understand the price system and how to structure their energy usage to maximise savings.
- More support and maintenance personnel training would be necessary as the controls and the whole system embrace more complexity.
- Advocates from local communities are crucial to success.
- Microgrid governance is crucial, as people prefer third party-led arrangements to deliver power.
- A project aimed at improving reliability for small groups of consumers can be harder to prove economically efficient, especially when reliability targets are averaged across the customer category.
- The microgrid needs competitive offerings (i.e., better than an alternative), willing customers, and a long-term operation ability.
- Participants' engagement with energy usage suggests they are likely to be open-minded about receiving more advice about how to get the most out of solar and batteries.
- Most people prefer a community battery to have their own battery considering investment, and maintenance cost and efficiency benefits.
- Engage the community early in the site selection process.
- Partnerships and early engagement are key to overcoming barriers to renewable energy adoption.
- Wind turbines being located so close to town caused some residents concern, but a well-designed community education process gained residents' support.
- Such projects can raise the awareness of how renewable energy sources and associated technology might provide reliable electricity generation in other communities with similar off-grid setups.
- In case there are high dust levels in a town, physical cleaning of the PV modules is needed. Therefore, educate the community members is important to keep the performance of solar PV systems high.
- Controlling the assets of third parties who are not required to cooperate is difficult.

9.1.5 Environmental Lessons Learnt

The collected lessons learnt considering the environmental aspects based on the national and international practices are listed as follows:

- The ability to eliminate intermittency issues caused by solar and wind generation in weak power systems gives renewable generation a chance to be used as the

primary source of electricity in many remote communities that do not or cannot access diesel fuel or other fossil fuels.
- If all islands are just taken into account for microgrid projects, it can save 1.5 GtonCO_2 per year (4% of the world's emissions).

9.2 Best Practices

The best practices in developing a framework to model, design and evaluate the microgrid projects depend on the requirements of stakeholders and the key application use case(s). Thus, involving the stakeholders and the main actors in this process can significantly shape the planning and operation of the microgrids to achieve the near and long-term goals. The judgment criteria in selecting the technologies, models, analysis frameworks and tools depend on the use case(s), main actors, location, and underlying data accuracy and availability. In addition, selecting the types of the models and methods can be done based on a trade-off between accuracy, simplicity, simulation burden and data availability. Hereby, reasonable assumptions should be made as securing 100% accuracy models and results are unrealistic expectations. According to the reviewed national and international microgrid projects (see Appendices A and B), the key process steps for designing and operating microgrid projects based on the best practices are shown in Fig. 9.1 and explained in detail as follows.

Step 1. Scenario(s) evaluation based on use cases and main actors

- Establish a stakeholder process to work with utilities and other interested stakeholders to select and refine the geographical location, applications, and requirements of designing, implementing, and operating the microgrids to meet the near and long-term goals.
- Involve the main actors in defining their needs and expectations to collect their opinions to frame the possible use cases.

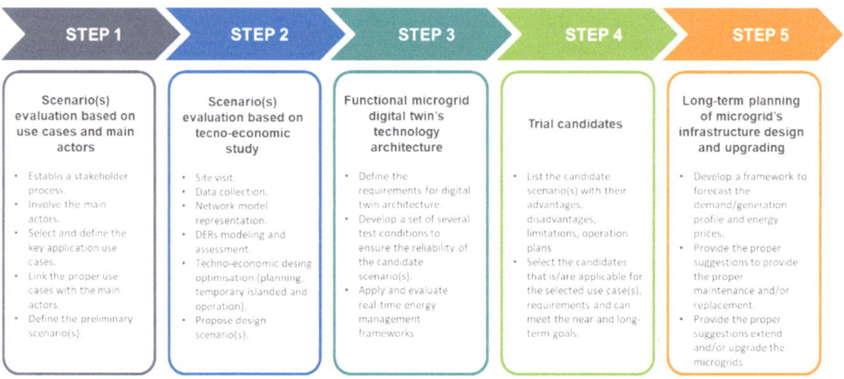

Fig. 9.1 Best practices of microgrid design and operation

- Select and define the use cases for designing and operating the microgrid with diverse stakeholders' input. The use cases should guide the development of modelling the microgrid components, techno-economic optimisation, operation, and its implementation. There are two main groups of use cases: (i) business use cases; (ii) and control and management use cases.
- Link the proper use cases with the main actors to build the interconnection between them in order to develop a proper business model to engage the actors in designing and operating processes. This will lead to participate the main actors in the response to the control and market signals.
- Define the preliminary scenario(s) based on the selected use case(s), main actors, and regulations.

Step 2. Scenario(s) evaluation based on techno-economic study

- Site visit of the candidate location to collect the required information and study the location in order to define the opportunities and challenges.
- Collect the relevant data and information, including the network information, demand data, meteorological data, tariff structure and standards.
- Understand the load behaviour, and the evaluate the efficiency of operating the loads based on the price efficiency index.
- Model the network representation, load, DER technologies and energy market structure based on the use case(s), and data availability.
- Propose reasonable assumptions if needed.
- Evaluate the potential of the DERs.
- Conduct a techno-economic planning optimisation. The selection of the objective functions(s) and the constraints can be done based on the use case(s) and the goals of the microgrid.
- Check the capability of the candidate design to operate in the islanded mode under several conditions and specific time period at high reliability. If the design could not meet the requirements, then the techno-economic planning optimisation should be modified and repeated until meeting the islanded operation requirements.
- Evaluate the operation of the candidate design under several conditions to ensure the optimal operation during all the operation modes at high availability and reliability at minimum cost. Here, if the design could not meet the requirements, the techno-economic planning optimisation should be modified and repeated again until meeting the operation requirements.
- Propose the design scenario(s) that can meet requirements and lead to achieving the near and long-term goals.

Step 3. Functional microgrid digital twin's technology architecture

- Define the requirements for building a functional microgrid digital twin technology architecture to evaluate the candidate scenario(s) in real-time emulation and validate the results based on the actual data and signals.
- Develop a set of several test conditions to ensure the reliability of the candidate scenario(s).

- Apply and evaluate real-time energy management frameworks to ensure the capability of the candidate scenario(s) to optimally operate and define the expected risks.

Step 4. Trial candidates

- List the candidate scenario(s) with their advantages, disadvantages, limitations, operation plans, risks, and benefit/cost assessment.
- Select the candidate(s) that is/are applicable for the selected use case(s), requirements and can meet the near and long-term goals.

Step 5. Long-term planning of microgrid's infrastructure design and upgrading

- Develop a framework to forecast the demand/generation profile and energy prices to ensure the optimal decision of the optimising the real-time operation, control actions and market signals.
- Provide the proper suggestions to provide the proper maintenance and/or replacement the microgrid components based on the operation status and degradation factors.
- Provide the proper suggestions extend and/or upgrade the microgrids to ensure meeting the near and long-term goals.

Summary

Consistent microgrid design and operation framework among DNSPs and microgrid operators in Australia is required. This consistent framework will lead for peer learning and experience exchange to overcome the challenges and limitations in order to increase the accuracy and functionality of designing and operating the microgrids over time. The consistent design and operation framework should be able to be implemented on different DER technologies, which may allow to increase the funding and implementation of microgrid projects around the country.

9.3 Standing Challenges and Barriers

Based on the reviewed national and international projects, assessing and implementation of microgrids are constrained by the limitations of:

- Microgrid research and/or projects are usually limited to fixed electrical boundaries with a single interface to the area's electricity systems. As a result, the multiple feeders in today's distribution grids do not participate in microgrid operations.

- Lack of consistent microgrid design and operation framework among DNSPs and microgrid operators.
- The majority of commercially available microgrid controllers are based on proprietary hardware or software and require dedicated high-speed communication links.
- With the inherent current limiting capability of DER inverters, the fault current under islanded conditions is much lower than that in grid-connected mode. This leads to ineffective or even non-operative system-level protection in microgrid.
- Access to data (i.e., smart meter data) and network information (i.e., network connectivity and impedances) are required for both model-based and data-driven approaches to ensure accurate results.
- MV and LV network model simplifications arising from limited access to data and network information can lead to grossly inaccurate results.
- Costs and benefits of coordinated aggregation of EVSE are not well understood. The potential solutions by combining traditional network voltage control approaches (i.e., OLTC transformers), strategically located sensors, and inverter volt-var, volt-watt functions, while practical, is not fully quantified.
- The equitable access of network resources to consumers DER remains an open challenge with steady-state or dynamic operating envelopes due to unavoidable locational diversity of DER connections.
- There are three major challenges when solving the OPF problem as follows:
  - Combination of systems: The best way to combine the different properties, optimisations, and costs into one formulation is still questionable.
  - Good solver: Integration of systems requires a good method to solve this problem, which can be referred to as a solver. This does not only apply to the optimisation problem itself, but also to the simulation within the optimisation, the computation of the power flow. Simulating more complex and integrated systems may be possible, depending on the solver.
  - More powerful hardware: Using a simple laptop or desktop, the simulation may take up to a week. The simulation process is accelerated significantly by powerful hardware, especially when it is tuned correctly to meet the problem.
- There is a lack of clarity on appropriate tariff frameworks for community microgrids. There is a need for activities that move the field further such as the development and trialling of appropriate and equitable microgrid tariffs and metering arrangements.
- Maintaining the grid stability with the DERs, especially those that are variable, is a challenge for most of the case studies. A phased integration approach can be followed to initially test the stability by integrating a small amount of DERs and start increasing the amount of DERs alongside with storage systems and advanced controls.
- Remote locations can be a logistic challenge for procuring, transporting, and installing the new technologies and getting the technical teams on-site either in the implementation phase or in case of any major issues.

- Lack of clarity in the correct legislative framework on the perception of microgrids with multi-user local ownership models.
- Lack of clarity in legislation for private electrical infrastructure in public roads.
- Lack of common technical national and international standards for microgrids.
- Lack of recognition of the social, economic, and environmental benefits of microgrids in the process of allocation of government infrastructure funding arrangements.
- There are some barriers in the wholesale energy market that limit some part of the implementation model.
- There are a lot of difficulties and challenges in controlling the microgrid assets that are owned by third part entity.

9.4 Research Roadmap

This section focuses on sharing three interrelated outcomes from the capability review of the national and international projects: barrier analysis, research priorities, and research milestone and activities. These research priorities help to frame the research roadmap, which identifies a number of milestones over the coming three years for each research priority. The research roadmap includes several research objectives with specific research activities.

9.4.1 Barrier Analysis

Real barriers have a contestant framework in the country for optimal and sustainable designing, temporary islanding and operating microgrids considering the technical, economic, environmental, and social aspects need to be addressed. Here, six main barriers are constrained the development of a contestant framework. Accordingly, six research priorities have been identified to overcome these barriers. Here, the six barriers associated with their research priorities are shown in Fig. 9.2.

9.4.2 Research Priorities

Several barriers are limiting the visibility, inadequate market integration, customer engagement and policy support that need to be addressed in the future research activities. Accordingly, six research priorities for targeted interventions have been identified through this capability review for the national and international practices as shown in Fig. 9.3.

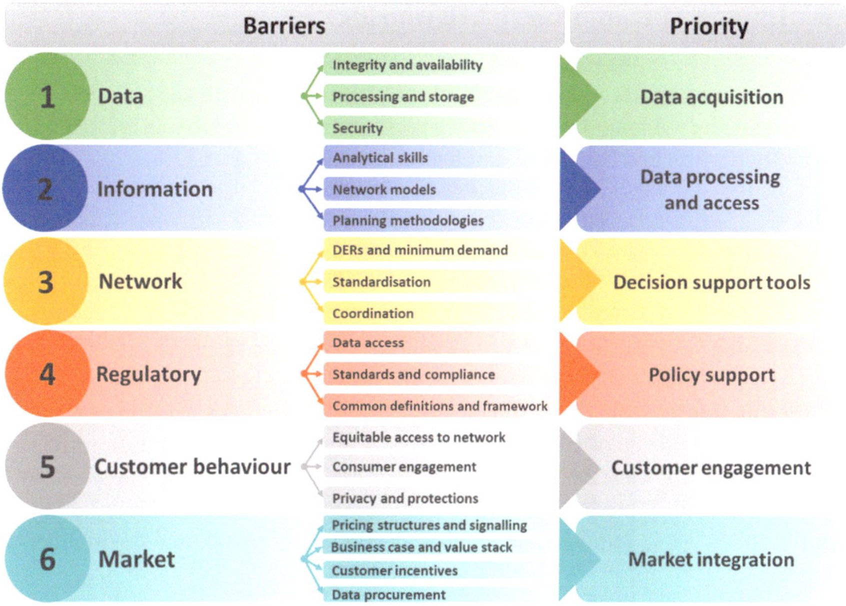

Fig. 9.2 Barriers' analysis and research priority matching

Fig. 9.3 Research priorities

9.5 Impact of Community Microgrids

Community microgrids will bring several social, environmental, and technical benefits for the wider communities. The structure of the framework is based on the CSIRO impact framework and the UK research excellence framework (REF).

9.5.1 Path of Impact and Impact Framework

A remote and regional microgrids can address the issues of increasing electricity price, aging infrastructure, and supply energy over wider geographical areas where the transmission of power over long transmission line is impractical/impossible. It can reduce the reliance of on diesel for remote communities in island. It introduces flexible options for increased energy reliability and independences. The market for remote microgrid is predicted to increase to $20 billion in Australia by 2024. A proper framework is required to quantify its economic, social, and environmental impacts.

9.5.2 Impact Categories and Key Performance Indicators

The impact of regional community microgrid is divided into three categories: (i) adoption new technology and skills developments impacts; (ii) technical (network) impacts; and (iii) economic and social impacts. In addition, this subsection focuses on identifying KPIs and metrics for three categories of outcomes. Each outcome type is subdivided into several sub-category outcomes, and there is at least one KPI and often several associated with each sub-category. Similarly, each KPI has at least one metric, though often several.

9.5.2.1 Adoption New Technology and Skills Development

Microgrids can accommodate a large amount of small and distributed renewable energy. As the design and operation of microgrids is highly due to its reliance on local infrastructure and community behaviours, it will offer a great opportunity to develop new knowledge and train the local community. Accordingly, the indicators and matrices for knowledge and technology diffusion outcomes are listed in Table 9.1.

Table 9.1 Categories and indicators for knowledge and technology diffusion outcomes

Category	Indicators	Metrics
New knowledge, and local skills	Improved capacity building through training and knowledge sharing	Number of people trained
	Upskilling in targeted and specialised areas	Number of skill sets identified
Attitudes	Increased community engagement and participation in community microgrids	Number of operational community microgrids
	Customer participation in wholesale markets	• Number of networks buying non-network support services • MW of non-network support traded
New technology	Increased integration of renewable energy and energy storage systems	Number of product/services sales, and number of retailers/aggregators offering services
	Increased uptake of tools for microgrid planning, design, and operation	Number of communities utilising the tools
	Model for utilising mix of new technology and existing assets	Number of network operator adopting models

9.5.2.2 Technical (Network) Impacts

Electricity networks in the twenty-first century will be completely different from those in the last century. Dispersed and distributed energy sources in microgrids will bring many opportunities and challenges for network operators. Smart community microgrids will help to facilitate the effective integration of cleaner, greener and more efficient DER units with improved visibility and customer engagement. Accordingly, the indicators and matrices for technical (network) impacts are listed in Table 9.2.

9.5.2.3 Economic and Social Impacts

The community microgrid will create new business opportunities for organisations and small companies in remote areas. It is important to note that integrating DER units in remote and regional microgrids will not only serve the primary purpose of energy injection but also contribute to reductions in energy losses and, thereby, reduce electricity bills and improve the national economy. By facilitating the robust interconnection of microgrids, it will provide new knowledge about ways of improving the efficiency and reliability of Australia's grid during natural disasters and maximising its advantages in critical power sectors. Accordingly, the indicators and matrices for technical (network) impacts are listed in Table 9.3.

Table 9.2 Categories and indicators for technical (network) Impacts

Category	Indicators	Metrics
DER integration	Improved visibility of microgrid networks	• Microgrid use cases • Percentage of visibility on network
	Better coordination to minimise export constraints	• Limits on power export • Average size of DER installations
	Increased DER integrations and hosting capacity	• Grid decentralisation ratio • Percentage of DERs on distribution networks
Operations of microgrid networks	Enhanced utilisation of microgrid infrastructure	• Percentage of utilisation • Annual load variation curve
	Operation of microgrid networks within operational range and standards	• Annual average breach hours/day • Evidence of voltage violations (over-voltage and under-voltage) • Metrices for power quality, for example voltage sags, swells, phase unbalance, transients, and harmonics)
Informed microgrid policy, standard and regulation	Influenced decision making/decision-makers	• Change of policy and guideline • Citations in key industry decision-making forums
	Informed changes in policy, market rules, legislation, regulations, or guidelines of community microgrids	
Equity, trust, and satisfaction of customers	Transparent and fair access to microgrid assets	• Agreement for network connections • Average number of customer complaint DER curtailment

Table 9.3 Categories and indicators for economic and social impacts

Category	Indicators	Metric
Economic	Lower energy cost for households and businesses	$ (Reduction in energy bill)
	Reduced cost of network	LCOE of DERs in microgrids
Economic/environmental	Improved reliability and resilience of power supply	Number and duration of power outage
Environmental	Lower CO_2 emissions	GHG equivalent
Social/economic	Creation of new jobs in microgrid sector	Full time equivalent (FTE)

9.6 Future Research Opportunities

The review and assessment carried out within this report indicate that there are many opportunities for future work to develop and improve the design, temporary islanding, and operation of microgrids. Some of the directions for future work can be listed as follows.

- Consistent microgrid design and operation framework

Further research is required to develop a consistent microgrid design and operation framework that is be able to be implemented on different DER technologies.

- Develop machine and deep learning techniques

Further research is required to develop a more accurate short-term forecasting to reduce the number of back-up supplies and their frequent operation.

- Power electronics interface for DERs

Microgrids can use smart transformers in the future, also known as solid-state transformers. With the smart transformer, an AC/DC/AC power conversion stage with a high-frequency transformer is used in place of the classical electromagnetic transformer in active MV to LV substations. By implementing power electronics converters into the MV/LV future substations, the converters can offer different types of services: voltage and reactive power control, power flow management, and the inclusion of a local energy storage device for microgrid islanding operation and support for black start strategies.

- Inverter-based battery energy systems

Advanced inverter-based battery energy systems will play an important role in island operation as they can help maintain the supply-demand balance in an islanded system alongside with the variable renewable energy generators. AEMO recommends prioritising the deployment of grid-forming the capabilities on grid-scale BESS, as this technology provides the capability to deliver firm, flexible energy behind the inverter.

- Power electronics dominated power system

Further research work is required to fully understand the operation of grid-forming resources under faults, both balanced and unbalanced. Additionally, the ability of the grid forming resources to energise transformers and the motor load is to be studied. Additionally, it yet to be understood if grid forming inverters would add any new modes of oscillation in a network. Although grid forming inverters PV system and battery energy storage systems are available, yet no feasible grid-forming inverters for wind turbines are available.

- EVSE modelling and representation with the proper regulation

More research is needed to study the possible ways of representing the behaviours of the EVs customers and propose a proper business model to avoid any legal issues and encourage them to contribute to improving the reliability and stability of microgrids.

- Innovative control architecture

Further research is required to improve the control architecture in microgrids to involve the negative sequence control strategies, which could be a solution helping the renewable energy microgrids flourish in terms of applicability and functionality. Accordingly, advanced control, operation and protection schemes can be analysed for robust operation of microgrid in low inertia islanding mode with variable penetration of renewable energy. This requires advanced stability analysis to observe the systems stability in different islanding conditions.

- Net-zero microgrids

There is a considerable GHG emission associated with today's microgrids, which largely use fossil-fuel generators. By replacing fossil-fuel generators with renewable energy, storage, and flexible loads, microgrids can play a significant role in the transition to a carbon-free distribution network. New dispatchable generation technologies that are 100% carbon-free can enhance this transition. They also provide a more dependable and sustainable power source, enhancing reliability and resilience. Microgrid decarbonisation entails three steps: (i) maximising the generation from renewable sources; (ii) managing storage and flexible loads; and (iii) introducing new clean energy sources, such as hydrogen fuel cells and small nuclear reactors. Such transitions bring complexities, especially in the area of energy management on both the utility and microgrid sides.

- Energy markets and examine new tariff structures

There is potential for microgrid customers, including owners of DER and smart appliances, to unlock and receive value from their flexibility. Future arrangements in two-sided energy markets that value the flexibility of energy use and demand resources can deliver benefits for all electricity system users. In addition, it is possible to extract even more value from the electricity network by supporting it with dynamic services, such as frequency and voltage support.

- Interaction between VPP/DER aggregators and microgrids

Further works are required to understand the interactions between virtual power plants and microgrids. It is not clear whether the consumer expectations and engagement pathways are the same or different for VPP's and microgrids. Further studies are needed to understand different business models for VPP's and microgrids and how can they be harmonised when both overlap on customers.

- A virtual net meeting market

Develop a model and evaluate the potential for trading energy services at the distribution level, such as demand response, aggregated generation, and peer-to-peer trading. Further research is required to establish a new market participant classification or classifications for 'New Energy Service Providers', including services like demand management, aggregations of and peer-to-peer energy trading.

- Engage customers and ensure information transparency

A customer-centred energy sector must focus on serving customer interests and customer goals. These interests and goals can only be clearly discerned and understood by effectively and openly engaging with customers themselves. In recent decades, policy makers and businesses have not always succeeded in achieving this. Further work is required to understand customer aspirations and expectations for the energy sector and enhance the level of trust between customers and service providers in the energy sector.

- Inclusion of thermal storage and super-capacitor in the microgrid

A hybrid energy storage system consisting of higher energy density and power density can provide desired performance of a microgrid to quickly ramp up output power in short-term and supply for longer duration during natural disasters. Advanced energy management systems and coordinated controllers are required to provide secured operation of microgrids.

- Future role of hydrogen in the energy mix

Australia is well placed to produce green hydrogen from solar and wind power due to its geography and abundant renewable energy sources. Hydrogen brings Australia great opportunities in varied areas, including export industry, power networks, gas networks, transport, and industrial processes. All of these hydrogen opportunities can raise some challenges for market activation. Although investment and scale-up will address many of these challenges, there is a subset of techno-economic analysis that should be addressed. Therefore, further research is required to study the future role of hydrogen in the energy mix and its potential application in the broader economy.

- Carbon-air battery as a next-generation energy storage system

Carbon-air batteries are promising technology which utilises carbon (C) as an energy source instead of hydrogen. It consists of a solid-oxide fuel and electrolysis cell where carbon generated via electrolysis of CO_2, is oxidised with air to produce energy. Carbon-air batteries can be charged using the energy generated by the renewable sources to reduce CO_2 to C. During the subsequent discharge phase, the C is oxidised to generate energy. This technology expected to have smaller size and higher efficiency than other hydrogen storage. Further research is required to evaluate the effectiveness of this technology technically and economically in microgrids.

Appendices

This report contains three main appendices that cover the capability reviews of 20 national microgrid projects in Appendix A, 10 international microgrid projects in Appendix B, and the national and international power grids and supporting technology standards in Appendix C.

Accordingly, Appendix A contains 9 tables that summarised the main features for planning and operating the microgrids around the country. In which, Table A.1 lists the information related to 20 national microgrid projects and 2 ongoing projects as well as it summarises the key application use cases. In addition, Table A.2 summarises the microgrid's technologies, modelling techniques and tools in the national microgrid projects. Also, Table A.3 reviews the power flow/state estimation techniques, analysis horizon and tools of the national microgrid's projects. Moreover, Table A.4 lists the optimisation problem definitions, functions, and constraints that employed in the national microgrid's projects. Furthermore, Table A.5 summarises the control, and EMS technologies and tools applied in the national microgrid's projects. The applied standards and guidelines for designing and operating microgrids in the national microgrid's projects are listed in Table A.6. Moreover, Table A.7 lists the digital twin technologies and data acquisition systems used in the national microgrid's projects. The implemented business models of the national microgrid's projects are mentioned in Table A.8. Finally, Table A.9 summarises the highlights, limitations, and recommendations based on the national microgrid's projects.

Also, Appendix B contains 9 tables that summarised the international practices in 10 microgrid projects, including the main features for modeling, planning, controlling, and operating the microgrids as well as the implemented digital twin and data acquisition systems considering the standards and guidelines. In this context, Table B.1 summarises the information related to 10 international microgrid projects as well as it lists the key application use cases. In addition, Table B.2 lists the microgrid's technologies, modelling techniques and tools in the international microgrid projects. Accordingly, the power flow/state estimation techniques, analysis horizon and tools of the international microgrid's projects are summarised in Table B.3.

I. A. Ibrahim et al., *Microgrids Design and Operation*, Power Systems,
https://doi.org/10.1007/978-981-96-4914-3

Also, Table B.4 lists the optimisation problem definitions, functions, and constraints that employed in the international microgrid's projects. Moreover, Table B.5 lists the control, and EMS technologies and tools applied in the international microgrid's projects. The applied standards and guidelines for designing and operating microgrids in the international microgrid's projects are summarised in Table B.6. Furthermore, Table B.7 summarises the digital twin technologies and data acquisition systems used in the international microgrid's projects. Also, Table B.8 lists the business models deployed in the international microgrid's projects. Finally, Table B.9 summarises the highlights, limitations, and recommendations based on the international microgrid's projects.

Finally, Appendix C covers the national and international power grids and supporting technology standards. In which, Tables C.1, C.2, C.3, C.4, C.5, C.6, C.7, C.8, C.9, C.10, C.11, C.12, C.13, C.14, and C.15 cover the national and international power grids standards, while the national and international supporting technology standards are summarised in Tables C.16, C.17, C.18, C.19, C.20, C.21, C.22, C.23, C.24, C.25, C.26, C.27, and C.28.

Appendix A

I. A. Ibrahim et al., *Microgrids Design and Operation*, Power Systems,
https://doi.org/10.1007/978-981-96-4914-3

Table A.1 Capability review of several national microgrid's projects

Code	Project	Year	City, state	Category	Type	Operation mode	Scope/objectives	Key use case(s)
NP1	Coral Bay Microgrid [1, 2]	2006	Coral Bay, WA	Residential, industrial	AC	Islanded	• High penetration of renewables • Lowest cost operation • Maximise wind farm output • Maintain operating limits of diesel plant • Reliable and stable power supply	• Microgrid provides Islanding mode • Balancing supply and demand on different timescales • Storage management • Supply-side management
NP2	King Island Renewable Energy Integration Project [3]	2010	King Island, TAS	Industrial, commercial and residential	Hybrid	Islanded	• Adding biodiesel tanks and infrastructure to existing diesel engine tanks to test the usage of biodiesel in diesel generators • Installation of two auxiliary diesel engines with flywheels, enabling for 100% renewable energy contribution • An integrated network of centrally managed smart meters and smart switches has been built to provide rapid responding demand side management	• Smart metering • Demand-side management • Storage management of energy in batteries to fulfill the supply and demand • Microgrid provides islanding mode
NP3	Marble Bar and Nullagine Microgrid [1]	2015	Marble Bar and Nullagine, WA	Residential, industrial	AC	Islanded	• Providing very high levels of solar energy penetration into the grid • Enhancing the system reliability • Minimise diesel consumption	• Microgrid provides islanding mode • Balancing supply and demand on different timescales • Storage management • Supply-side management

(continued)

Table A.1 (continued)

Code	Project	Year	City, state	Category	Type	Operation mode	Scope/objectives	Key use case(s)
NP4	Delivering higher renewable penetration in new land and housing developments through edge-of-grid microgrids [4]	2017	Hunter Valley, NSW	Residential and commercial	AC	Islanded	• Provide generation, network and retail services • Optimal coordination of dispatchable generation, PV, energy storage and in some cases residential load	• Controllable power range determination on different timescales • Storage management • Protection and restoration • Smart metering • Data management
NP5	Moreland Micro-Grid Investigation [5]	2017	Melbourne, VIC	Residential	AC	Grid-connected	• Deploying more solar energy than usual • The reduction of household participation costs • Optimising the demand profile in network terms	• Balancing supply and demand on different timescales • Demand-side management • Storage management • Smart metering
NP6	DeGrussa Solar Project [6]	2017	Meekatharra, WA	Commercial	Hybrid	Islanded	• The project's goal is to increase awareness and trust in the use of renewable energy to power off-grid mine sites	• Storage management of energy in batteries to fulfill the supply and demand • Microgrid provides islanding mode

(continued)

Table A.1 (continued)

Code	Project	Year	City, state	Category	Type	Operation mode	Scope/objectives	Key use case(s)
NP7	Karratha Airport Solar Plant [7, 8]	2018	Karratha Airport, WA	Commercial	AC	Grid-connected	• Investigate the operation of the 1 MW Karratha Airport solar plant • Evaluate the performance of the cloud prediction (CPT) and battery technology both used to manage the electricity flow and smooth PV output variations • Improve the competitiveness of the main grid and beyond by showing the role of renewable DERs integrated with CPT	• Forecasting energy production and consumption on different timescales • Storage management • Microgrid Operator sells/buys energy on external market • Data management
NP8	Latrobe Valley Microgrid Feasibility Study [9]	2019	Latrobe Valley, VIC	Residential, commercial, and farm (virtual microgrid)	AC	Grid-connected	• Assessing community perceptions of the value of peer-to-peer energy trading in the Latrobe Valley • Analysing the potential economic value of the Local Energy Market (LEM) for both consumers and prosumers • Evaluating the regulatory barriers to implementing a LEM and possible ways of addressing them	• Microgrid provides an open trading and communication platform for internal trading • Matching demand and supply on different timescales (renewable generation) • Demand side management • Smart metering
NP9	ATCO Hydrogen Microgrid [10, 11]	2019	Jandakot, WA	Commercial	Hybrid	Grid-connected and islanded	• It comprises the design and building of a commercial size Hybrid Energy System (HES) micro-grid • It demonstrates how a commercial scale microgrid integrating solar, battery and gas • Generation can address availability and variability issues • It introduces on-site hydrogen generation from excess solar energy	• Microgrid provides islanding mode • Storage management of hydrogen • Smart metering

(continued)

Table A.1 (continued)

Code	Project	Year	City, state	Category	Type	Operation mode	Scope/objectives	Key use case(s)
NP10	Onslow Distributed Energy Resources Management System (DERMS) [12]	2019	Onslow, WA	Residential	AC	Grid-connected	• A 1 MVA solar PV system adjacent to the new Onslow power station • The Onslow DER project connects traditional energy sources with distributed solar and battery solutions to maximise the amount of renewable energy • DERMS balances supply and demand, keep the electricity system stable and provide long-term economic value	• Protection and restoration • Supply-side management • Planning • Storage management
NP11	Indra Monash Smart City [13]	2019	Clayton, VIC	Residential	AC	Grid-connected	• The assets will be monitored in real time, allowing for visibility and control of the low-voltage network • The assets will be optimised to ensure an efficient and reliable supply of electricity • In the third and final phase, a transactive market will be implemented, allowing each building to buy and sell electricity and optimise usage in response to pricing signals	• Demand side management • Storage management • Smart metering • Planning • Microgrid provides islanding mode
NP12	Garden Island Microgrid Project [14]	2020	Garden Island, WA	Military	AC	Grid-connected and islanded	• The first wave energy integrated microgrid in the world, generating both electricity and desalinated water • It comprises the design, construction, and commissioning of a 2 MW solar PV system with a 2 MW/0.5 MWh battery energy storage system • It includes the potential of incorporating wave energy into the microgrid, as well as the desalination plant	• Balancing supply and demand on different timescales • Demand-side management • Storage management of energy in batteries to fulfill the supply and demand

(continued)

Table A.1 (continued)

Code	Project	Year	City, state	Category	Type	Operation mode	Scope/objectives	Key use case(s)
NP13	Horizon Power Carnarvon Distributed Energy Resources (DER) Trials [15, 16]	2020	Carnarvon, WA	Residential and commercial	AC	Islanded	• Resolve the technical, operational, and transitional barriers to a high penetration DER business in remote networks • How to manage DER in a microgrid with respect to the power quality and value exchange with customers • Build capabilities in the management and optimisation of renewable generation	• Supply-side management • Active power voltage control • Enhance local power quality • Smart metering
NP14	Study into the feasibility of a microgrid at Kalbarri [17]	2021	Kalbarri, WA	Township	AC	Islanded and grid-connected	• A more reliable power supply • Enhance Western Power's understanding of microgrids and their interaction with the network • Developing a replicable and adaptable model for regional power supply • Enhanced resilience achieved through the microgrid's islanding capabilities	• Protection and restoration (mitigate power reliability issues) • Microgrid provides islanding mode
NP15	Solar and Storage Trial at Alkimos Beach Residential Development [18]	2021	Alkimos Beach, WA	Residential	AC	Grid-connected	• Demonstrating the feasibility of a new model of energy service that virtually connects individual household PV systems with a community-scale energy storage system • Investigating an energy smart home package • New electricity retail products and services • New metering and data platforms	• Microgrid provides an open trading and communication platform for internal trading • Demand-side management • Storage management • Data management • Smart metering

(continued)

Table A.1 (continued)

Code	Project	Year	City, state	Category	Type	Operation mode	Scope/objectives	Key use case(s)
NP16	Yackandandah Microgrid Feasibility Study [19, 20]	2021	Yackandandah, VIC	Township	AC	Grid-connected	• To determine the best power solution for achieving 100% renewable energy • To identify potential sites and develop solutions/installations for renewable energy • To determine capital costs, operating costs, and energy generation and storage capacities for each site-specific option • Evaluation of the financial and environmental feasibility of each option • Identifying potential community ownership models for the options identified (community-scale/grid-facing assets)	• Trading energy with external markets • Selling balancing and ancillary services • Demand-side management • Supply-side management • Storage management
NP17	Increasing the Uptake of Solar PV in Strata Residential Developments [21]	2021	White Gum Valley, City of Fremantle, WA	Residential	AC	Grid-connected	• Determine the key factors influencing market adoption of solar PV and storage solutions in a medium-scale urban strata development • Create a governance model and microgrid design for solar PV, battery storage, and metering solutions • Examine the costs, benefits, and risks of the chosen governance model from the standpoint of all affected stakeholders	• Supply-side management • Storage management • Protection and restoration • Smart metering • Planning
NP18	Advanced Energy Resources Wind, Solar and Battery Project [22]	2021	Port Gregory, WA	Industrial	Hybrid	Grid-connected and islanded	• The objective of the project is to build a new grid connection design that overcomes the challenges of connecting large amounts of renewable energy to weak grids and provides backup power	• Microgrid provides islanding mode

(continued)

Table A.1 (continued)

Code	Project	Year	City, state	Category	Type	Operation mode	Scope/objectives	Key use case(s)
NP19	Lord Howe Island Hybrid Renewable Energy System [23]	2021	Lord Howe Island, NSW	Residential	AC	Islanded	• Improve the self-sufficiency by reducing the Island's reliance on imported diesel fuel (under one third) • Lower long-term cost of energy production and avoid possible diesel price shock • Develop innovative energy management technologies • Protecting the tourism values by reducing the impacts of the use of the diesel fuels, including atmospheric pollution, traffic, and noise	• State analysis and sub-sequent actions for providing operation in islanding mode • Data management • Monitoring • Storage management
NP20	Flinders Island Hybrid Energy Hub [24]	2021	Flinders Island, TAS	Commercial	AC	Islanded	• Reduce the cost of operating high-penetration hybrid systems • Develop an implementation process with minimal impact on an existing power station • Reduce the capital cost up to 30% over the next 5 years	• State analysis and sub-sequent actions for providing ancillary services • Supply-side management • Planning
NP21	Horizon Power Denham Hydrogen Demonstration [25]	2020–present	Denham, WA	On going project				
NP22	Agnew Renewable Energy Microgrid [25]	2019–present	Leinster, WA	On going project				

Table A.2 National capability review of microgrid's technologies, modelling techniques and tools

Code	Network topology	Network data	Technology	Technology data	Time horizon	Modelling technique	Modelling tool
NP1	N/A	N/A	Wind turbine (3x 200 kW)	• Data recording at Coral Bay, 2008 • Vergnet tilt-up technology	N/A	N/A	HOMER
			Diesel generator (7x 320 kW)	Variable-speed low-load generator	N/A	N/A	HOMER
			Normal load	Data recording at Coral Bay, 2008	N/A	N/A	HOMER
			Flywheels storage (500 kW)	• Kinetic energy storage device • Modular inverter design to change kVA rating • Highly dynamic performance with large power ratings	N/A	N/A	HOMER
NP2	N/A	Information of interconnection of renewable energy sources, enabling technologies, thermal generation, and smart grid	PV (470 kW)	N/A	Hourly	N/A	N/A
			Wind turbine (2.5 MW)	N/A	Hourly	N/A	N/A
			Diesel generator (6 MW)	N/A	Hourly	N/A	N/A
			Electrical battery (3 MW/ 1.5 MWh)	N/A	Daily	N/A	N/A
			Flywheel storage (1 MVA)	N/A	Daily	N/A	N/A
			EVs charging station	Smart meter data	Hourly	N/A	N/A
			Controllable load (smart grid-demand side management)	Smart meter data	Hourly	N/A	N/A
NP3	N/A	N/A	Solar PV (508 kW)	Single axis tracking solar farms generation	N/A	N/A	N/A
			Diesel generator (2280 KW)	N/A	N/A	N/A	N/A

(continued)

Table A.2 (continued)

Code	Network topology	Network data	Technology	Technology data	Time horizon	Modelling technique	Modelling tool
			Normal load	N/A	N/A	N/A	N/A
			Flywheels storage (1000 kW)	Kinetic energy storage device	N/A	N/A	N/A
NP4	Radial	N/A	• Virtual solar 3–6 kW on 95% of homes • Solar Farm—4 MW (central)	• Extreme weather events • PV module cost curve	1–15 min period	Machine Learning and general PV model	• HOMER Pro • Australian PV institute tool
			• Gas generators • Diesel generators	N/A	N/A	N/A	HOMER Pro
			Electrical battery	Battery cost curve	N/A	N/A	HOMER Pro
			EV charging station	Seasonal peak demands	N/A	N/A	HOMER Pro
			Controllable loads (air-condoning, smart hot water, and heat recovery equipment)	• Extreme weather events • Seasonal peak demands	Hourly	N/A	HOMER Pro
NP5	N/A	N/A	PV	• Solar PV generation (kWh p.a.) • Peak generation • System size • Geographic location	Annual and monthly	N/A	N/A

(continued)

Table A.2 (continued)

Code	Network topology	Network data	Technology	Technology data	Time horizon	Modelling technique	Modelling tool
			Normal load	• Energy use (kWh) • Peak demand (kW) • Derived from smart meter data	Daily and annual	• The four types of households are: • Single/couple: small household, at work during the day • Family Type 1: large household, at work/school during the day • Family Type 2: large household, at home during the day • Retirees: small household, at home during the day	N/A
			Battery storage	• Capacity of the battery (kWh) • Depth of discharge • Round trip efficiency	N/A	Modelled as being consistent	N/A
NP6	Radial	Site map of project interfaces	PV (10.6 MW)	Solar radiation	Daily	N/A	N/A
			Diesel generators (1 MW)	N/A	N/A	N/A	N/A
			Electrical battery (6 MWh)	N/A	N/A	N/A	N/A
			Normal load (mining load)	N/A	N/A	N/A	N/A
NP7	N/A	N/A	PV (1055 kWp)	PV output ramp-up and ramp-down rates	1-s temporal resolution	Cloud-prediction technology	N/A
			Electrical battery (234 kW)—grid-support battery	• Battery output/input • SOC and DOD	1-s temporal resolution	N/A	N/A

(continued)

Table A.2 (continued)

Code	Network topology	Network data	Technology	Technology data	Time horizon	Modelling technique	Modelling tool
			• Load (500 GWh/ yearly): • Large loads • Interruptible loads (chillers, pumping, fans etc.)	Load curve	1-s temporal resolution	N/A	N/A
NP8	N/A	N/A	PV	• Installed capacity • Peak generation • Total annual generation	Daily and annual	• Existing solar systems and proposed solar settings (future installation) are modelled	PSS\E
			Normal load	• Average load • Peak load • Total energy usage (peak and off-peak)	30-min basis	• Buyer/seller Counterfactual ($) and Cost of purchases/sale on LEM ($) are modelled and compared	PSS\E
NP9	N/A	Single line diagram of microgrid network	PV (3×100 MW)	N/A	N/A	N/A	N/A
			Natural gas fired generators (200 kW)	N/A	N/A	N/A	N/A
			Fuel cell (1.1 kW)	N/A	N/A	N/A	N/A
			Electrical battery	N/A	N/A	N/A	N/A
			Hydrogen electrolyser	N/A	N/A	N/A	N/A
NP10	Radial	N/A	• Centralised PV system (1 MVA) • 261 residential PV systems (1341 kW) • 8 commercial PV systems (640 kW)	N/A	N/A	N/A	N/A

(continued)

Table A.2 (continued)

Code	Network topology	Network data	Technology	Technology data	Time horizon	Modelling technique	Modelling tool
			• 2 utility power station BEES (1 MWh and 500 kWh) • 19 residential energy system (190 kWh) • 8 commercial smoothing storage (361 kWh)	N/A	N/A	N/A	N/A
			• Centralised power station: • Gas-fired modular power station • Diesel generator for black start	N/A	N/A	N/A	N/A
			Loads	N/A	N/A	N/A	N/A
NP11	Urban	Single line diagram of microgrid network	PV (1 MW)	N/A	N/A	N/A	N/A
			Electrical battery (1 MWh)	N/A	N/A	N/A	N/A
			EV charging station (22 kW)	N/A	N/A	N/A	N/A
			Flexible building loads	N/A	N/A	N/A	N/A
NP12	N/A	N/A	PV (2 MW)	N/A	N/A	Machine Learning	N/A
			Electrical battery (2 MW/ 0.5 MWh)	N/A	N/A	N/A	N/A
			Normal load (miliary load)	N/A	N/A	N/A	N/A

(continued)

Table A.2 (continued)

Code	Network topology	Network data	Technology	Technology data	Time horizon	Modelling technique	Modelling tool
NP13	Rural (remote area)	Energy consumption, current, phase and line voltages, frequency, active and reactive power in all phases	PV	Maximum aggregated PV output, PV fluctuation factor, diversity factor, PV array rating, cloud Shading	5–15 min	• Machine learning models to develop detailed forecasts of renewable energy generation • K-means clustering model for low-voltage feeders • Real-time decision making for DER control • Analytical (the PV fluctuation factor)	Digsilent power factory
			Electrical battery		5–15 min	N/A	Digsilent power factory
			Normal loads		5–15 min	N/A	Digsilent power factory
NP14	Radial (rural long—150 km, 33 kV)	N/A	PV (5 MW)	PV generation	30-min basis	The output of an existing solar farm scaled to represent a 5 MW solar farm in Kalbarri	PROPHET
			Solar-thermal (1 MW)	Output generation and performance	N/A	Modelled as a dispatchable unit	PROPHET
			Wind	Wind generation	30-min basis	N/A	PROPHET
			Normal load	Load demand	30-min basis	• The load profile is assumed to follow a similar trajectory to historical data obtained between 1st July 2014 and 1st July 2015 • In order to account for additional load, the demand was scaled up by 23%	PROPHET

(continued)

Table A.2 (continued)

Code	Network topology	Network data	Technology	Technology data	Time horizon	Modelling technique	Modelling tool
			Battery storage	Capacity of the battery	N/A	The battery size is estimated based on a range of potential fault scenarios generated by PROPHET	PROPHET
NP15	N/A	N/A	PV—Rooftop (1.5–3.11 KW)	N/A	N/A	N/A	N/A
			Solar hot water system	N/A	N/A	N/A	N/A
			Normal load	N/A	N/A	N/A	N/A
			Battery storage	'Set and forget' charge and discharge cycle	N/A	• Virtual energy storage for $11 a month • Solar PV owners will receive solar credits instead of selling excess electricity to the grid • Energy consumption between 4 pm and midnight will be offset by these credits	N/A
NP16	Radial	N/A	Solar PV	Solar generation (existing rooftop PV)—generation data from the distribution network operator AusNet Services	N/A	The AEMO forecasts for increased solar PV are applied to solar PV growth	N/A
			Solar + battery	Fixed panels	N/A	N/A	N/A
			Solar + pumped hydro	• Bi-facial single axis tracking panels • Pumped hydro energy storage	N/A	N/A	N/A

(continued)

Table A.2 (continued)

Code	Network topology	Network data	Technology	Technology data	Time horizon	Modelling technique	Modelling tool
			Normal load	Total consumption (AusNet interval data)	At 15-min intervals	• Total demand estimation for last three years by adding the imported power (grid) to the locally generated power • To forecast future demand/consumption growth, the AEMO forecast increases are combined with an additional 3% local development assumption	N/A
NP17	Medium scale urban	Block diagram of microgrid networks containing PV system, battery energy storage system, energy peters and protection equipment	PV—rooftop (9 kW)	Solar radiation, humidity and temperature	Daily	N/A	N/A
			Electrical battery (150 kWh)	Charging and discharging data	Daily	N/A	N/A
			EV charging station	Charging data	N/A	N/A	N/A
			Normal loads (residential)	Load curves	Daily	N/A	N/A
NP18	Rural	N/A	PV (1 MW)	N/A	N/A	N/A	N/A
			Wind turbine (2.5 MW)	N/A	N/A	N/A	N/A
			Electrical battery (2 MW/ 0.5 MWh)	N/A	N/A	N/A	N/A
			Normal load (mine loads)	N/A	N/A	N/A	N/A
NP19	N/A	• Solar farm technical spec. • Automatic operating modes • Single line diagram (SLD) • Protection SLD • PV system layout • 415 V Powerhouse Main Distribution Board (MDB)	PV (1328.4 kWp DC/ 1000 KVA)	N/A	N/A	The PVsyst model is updated with actual measured meteorological data. This model is then imported into the Homer model which is updated with the actual load data over the period of 12 months (Same for both models)	PVsyst and HOMER
			Diesel generator (3× 280 kW)	N/A	N/A		PVsyst and HOMER
			Electrical battery (1 MW/ 3.7 MWh)	N/A	N/A		PVsyst and HOMER

(continued)

Table A.2 (continued)

Code	Network topology	Network data	Technology	Technology data	Time horizon	Modelling technique	Modelling tool
			Loads	N/A	N/A		PVsyst and HOMER
NP20	N/A	Main bus: 11 kV	PV (175 kW)	N/A	N/A	N/A	N/A
			Wind turbine (900 kW and privately owned 300 kW)	N/A	N/A	N/A	N/A
			Dynamic resistor (1 MW)	N/A	N/A	N/A	N/A
			Diesel-UPS (850 kVA)	N/A	N/A	N/A	N/A
			BESS (750 kW/266 kWh and 500 kW continuous rated)	N/A	N/A	N/A	N/A

Table A.3 Power flow/state estimation techniques, analysis horizon and tools of the national microgrid's projects

Code	Power flow/state estimation	Balanced/unbalanced	Steady-state/dynamic	Snapshot/time-series	Steady-state PF tool	Dynamic PF tool
NP1	Power flow	N/A	Steady-state	N/A	PowerFactory—DIgSILENT	N/A
NP2	N/A	N/A	N/A	N/A	N/A	N/A
NP3	N/A	N/A	N/A	N/A	N/A	N/A
NP4	Power flow	Balanced	N/A	Time-series	N/A	N/A
NP5	N/A	N/A	N/A	N/A	N/A	N/A
NP6	N/A	N/A	N/A	N/A	N/A	N/A
NP7	N/A	N/A	N/A	N/A	N/A	N/A
NP8	Power flow	Balanced	Both	Time-series	PSS/E	PSS/E
NP9	N/A	N/A	N/A	N/A	N/A	N/A
NP10	N/A	N/A	N/A	N/A	N/A	N/A
NP11	N/A	N/A	N/A	N/A	N/A	N/A
NP12	N/A	N/A	N/A	N/A	N/A	N/A
NP13	Power flow	Unbalanced	Steady-state	Time-series	N/A	PowerFactory—DIgSILENT
NP14	Power flow	N/A	Steady-state	Time-series	N/A	N/A
NP15	N/A	N/A	N/A	N/A	N/A	N/A
NP16	N/A	N/A	N/A	N/A	N/A	N/A
NP17	Power flow	Balanced	Steady-state	Time-series	N/A	N/A
NP18	N/A	N/A	N/A	N/A	N/A	N/A
NP19	N/A	N/A	N/A	N/A	N/A	N/A
NP20	N/A	N/A	N/A	N/A	N/A	N/A

Table A.4 Optimisation problem definition, functions, and constraints of the national microgrid's projects

Code	Single- or multi-objective/ type	Optimisation data	Objective function(s)	Technical constraint(s)	Values	Economic constraint(s)	Values
NP1	Single-objective/ techno-economic	• Load consumption • Step load response of diesel generator • Estimated wind generation • Diesel generator characteristics • Flywheels storage characteristics	Maximising Return on Investment (ROI) and fuel savings	N/A	N/A	N/A	N/A
NP2	Single-objective/ techno-economic	• Cost of producing power from diesel generators • Utility revenues	Minimising the carbon dioxide emission while improving the reliability of the system	N/A	N/A	N/A	N/A
NP3	N/A	N/A	N/A	N/A	N/A	N/A	N/A
NP4	Single objective/ techno-economic	Technology price curves, energy cost indexing	Minimising production costs (The optimal generation mix was developed using a long run marginal cost approach)	Frequency and voltage stability	N/A	N/A	N/A
NP5	Multi-objectives/ techno-economic	• Technical data: (i) amount of energy (kWh) used by a household; (ii) peak demand (iii) annual solar PV generation (kWh p.a.); and (iv) battery storage characteristics • Economic data: (i) capital costs of solar PV, battery, control system, and networking infostructure; (ii) operating and maintenance costs; (iii) administrative costs	• Minimising infrastructure costs • Minimising energy cost for consumers • Minimising peak event • Maximising capacity of rooftop solar • Grid import/export • Battery charge/discharge • Battery SOC	N/A	N/A	N/A	N/A

(continued)

Table A.4 (continued)

Code	Single- or multi-objective/ type	Optimisation data	Objective function(s)	Technical constraint(s)	Values	Economic constraint(s)	Values
NP6	Multi-objective/ techno-economic	• Load profile of mining site • Solar generation profile under sunny and cloudy conditions	• Maximising the power quality • Maximising the reliability of the system • Reducing the diesel generator run time	Minimum diesel generator loading	50%	N/A	N/A
NP7	Single-objective/ techno-economic	• Cloud predicted data • PV output data • Battery information	Maximising the PV output power	The ratio of the instantaneous PV ramp rate and PV nominal AC power rating	• Minimum ramp rate in p.u. = −0.083 • Maximum ramp rate in p.u. = 0.167	N/A	N/A
NP8	Single-objective/ techno-economic	Historical data on usage, solar generation profiles, and tariffs for each LVM participant	• Maximising profits of market participants • Supply and demand balance between solar exports and demand for every 30 min • Economic outcomes available to participants	N/A	N/A	N/A	N/A
NP9	Single-objective/ techno-economic	Carbon dioxide emission profile by burning natural gas	Maximising the production of hydrogen to minimise the carbon dioxide emissions by the user while maintaining an affordable energy service	N/A	N/A	N/A	N/A
NP10	Single-objective/ techno-economic	N/A	Minimising energy costs	N/A	N/A	N/A	N/A
NP11	Single-objective/ techno-economic	Internal and external market data	Minimising the energy cost of the microgrid network	N/A	N/A	N/A	N/A

(continued)

Table A.4 (continued)

Code	Single- or multi-objective/ type	Optimisation data	Objective function(s)	Technical constraint(s)	Values	Economic constraint(s)	Values
NP12	Single-objective/ techno-economic	N/A	Maximising the penetration of variable renewable energy resources to meet the requirement of Paris agreement (COP 21)	N/A	N/A	N/A	N/A
NP13	Single-objective/ techno-economic	N/A	Maximising aggregated PV output (considering cloud events in the analysis period)	Voltage limits	240 V ±6%	N/A	N/A
NP14	Single-objective/ techno-economic	• Mean Time to Fail (MTTF) • Mean Time to Repair (MTTR) • Forecasted demand and generation • Failure rate of various elements in a Monte Carlo environment	• Maximising the reliability of supply: • Estimation of the battery capacity • Unserved energy	The network is assumed to have no constraints (the network link between Kalbarri and the SWIS)	N/A	N/A	N/A
NP15	Single-objective/ techno-economic	'Set and forget' setup for community battery: Charging during low and discharging during high system demand	Optimising market costs and supporting the electricity system as a priority	N/A	N/A	N/A	N/A

(continued)

Table A.4 (continued)

Code	Single- or multi-objective/ type	Optimisation data	Objective function(s)	Technical constraint(s)	Values	Economic constraint(s)	Values
NP16	Single-objective/ techno-economic	• The capital/establishment costs • Ongoing operating costs • Current and forecasted electricity demand/ consumption • Estimation of local PV generation • Site identification/section attributes • Feeder information (lines, transformers, etc.)	Maximising the capacity of renewable energy	N/A	N/A	N/A	N/A
NP17	Multi-objective/ techno-economic	• Capital and operating and cost of PV and battery energy management system • Average daily load profile	• Minimising the capital cost of PV and battery energy management system • Minimising the cost of electricity for consumers	Availability of roof space to cover the number of panels required for energy needs of the resident	N/A	N/A	N/A
NP18	Single-objective/ techno-economic	N/A	Maximising the penetration of customer side renewable power generation	N/A	N/A	N/A	N/A
NP19	Single-objective/ techno-economic	N/A	Maximising diesel fuel savings (reducing fuel costs)	N/A	N/A	N/A	N/A
NP20	Single-objective/ techno-economic	N/A	Minimising the operation cost	N/A	N/A	N/A	N/A

Table A.5 Control and EMS technologies and tools applied in the national microgrid's projects

Code	Control level(s)	Local controllers/ technology	Secondary and tertiary control	EMS type (day-ahead/online)	Monitored values	Operation optimisation function	Operation optimisation tool	Service provider
NP1	Primary and secondary	• MGC600-G controller: • Generator, feeder, wind turbine	Distributed	N/A	N/A	• Ensures a higher overall plant availability • Communicate with each other • Regular updates are made to the information • Firmware includes corrective measures	HOMER	N/A
NP2	Primary	DER, thermal generation, enabling technologies and demand side-management control in islanded	N/A	Online	• PV power • Wind power • Battery power • Diesel power • Flywheel power • Dynamic resistor power • Total customer demand	N/A	N/A	Display of real time monitored values on mobile application
NP3	N/A	N/A	N/A	N/A	N/A	N/A	N/A	N/A
NP4	Primary and secondary	• Microgrid management system • Smart controllers at each customer connection	Decentralised	N/A	A variety of air-conditioning, hot water, heat pump and EV charging station	N/A	• Hybrid Plant Optimiser (HPO): optimises the operation of gen. units • Hybrid Plant Controller (HPC): manage the dispatch of different gen	• PV output and Load forecast • (Anticipated peak, anticipated hot weather)
NP5	N/A	N/A	N/A	N/A	N/A	N/A	N/A	N/A
NP6	Primary	DER, load	N/A	N/A	N/A	N/A	N/A	N/A

(continued)

Table A.5 (continued)

Code	Control level(s)	Local controllers/ technology	Secondary and tertiary control	EMS type (day-ahead/online)	Monitored values	Operation optimisation function	Operation optimisation tool	Service provider
NP7	Primary	• A ramp rate (PV output) control system • Adjusting battery input/output by monitoring clouds in the sky to smooth generation supply	N/A	Online	• PV arrays output • PV output ramping rates • DOD based on SoC values	N/A	N/A	Cloud images in the sky (Image recognition)
NP8	Primary and secondary	DER, load	Distributed	N/A	• Load and generation profiles, generating efficiency, carbon intensity and • Location-based pricing	Utilising blockchain technology to balance demand and supply and maximise load shifting	• TAG-e (hybrid meter/computer device) has two layers of functionality: • The trading of energy between devices via a transactive layer • The control layer contains control logic connecting the transactive layer with the device outcomes	• A mobile application for the setting of: • Price sensitivity • Energy preferences, and various devices' demand response capabilities
NP9	Primary and secondary	DER Control in Islanded and grid connected mode	Centralised control level	N/A	• PV power • Battery storage • Hydrogen production • Hydrogen Storage • Hydrogen usage	N/A	N/A	• AEMO for energy market data • BOM for weather data

(continued)

Table A.5 (continued)

Code	Control level(s)	Local controllers/ technology	Secondary and tertiary control	EMS type (day-ahead/online)	Monitored values	Operation optimisation function	Operation optimisation tool	Service provider
NP10	Primary and secondary	DER management system (an intelligent software-based system)	N/A	• Online • (Through Secure Gateway Device (SGD))		N/A	N/A	N/A
NP11	Primary and secondary	DER, load	Centralised	Online	Voltage Frequency Power quality	Maximise the power quality	N/A	Monitoring portal is created that allows users to visualise on screen graphical representations of the smart energy city real-time data, events and warnings, as well as historical data
NP12	N/A	N/A	N/A	N/A	N/A	N/A	N/A	N/A
NP13	Primary	PV and energy storage	N/A	N/A	Voltage (High or low voltage events)	N/A	N/A	Weather-energy-related dataset
NP14	N/A	N/A	N/A	N/A	N/A	N/A	N/A	N/A
NP15	Primary and secondary	Household air conditioners/direct load control technology	Centralised	N/A	N/A	• Reducing electricity demand during peak demand periods • To provide savings to households	N/A	N/A

(continued)

Table A.5 (continued)

Code	Control level(s)	Local controllers/ technology	Secondary and tertiary control	EMS type (day-ahead/online)	Monitored values	Operation optimisation function	Operation optimisation tool	Service provider
NP16	Primary and secondary	Household smart energy controller (with CTs on inverter, supply, and various household circuits)	Centralised	N/A	Household's demand, generation and storage status	• Voltage control capability • Reducing peak demand • Dynamically support the network • Flexible export control (smart exports)	• Mondo Ubi smart energy controller: • Connected to the Internet via the home Wi-Fi network or 3G/4G • Shared data with the microgrid operator (Mondo) • Instructions send via the 'cloud' to a household	N/A
NP17	Primary and secondary	DER, House services, studio, and apartment and	Distributed	Online	• Grid energy meter • EV charging meter • Load meter • PV/Battery energy storage system meter	N/A	N/A	• AEMO for forecasted energy market data • Home energy monitoring system on mobile application of individual user
NP18	N/A	N/A	N/A	N/A	N/A	N/A	N/A	N/A
NP19	Primary and secondary	Tesla microgrid controller (Managing the Tesla Powerpack inverters, SMA solar string inverters, and diesel generator controllers)	Centralised	Online/Automatic	• Real-time performance monitoring of: • The PV inverters • Storage system • Diesel generator • HMI	Predicting target renewable fraction for the system	HOMER	N/A
NP20	Primary	PLC-based control System (HMI is employed for updating the control code)	N/A	N/A	N/A	N/A	N/A	An online web-based dashboard

Table A.6 Standards and guidelines for designing and operating microgrids in the national microgrid's projects

Code	Microgrid specific standards	Power grids		Supporting technologies	
		Category	Standard	Technology	Standard
NP1	N/A	N/A	N/A	N/A	N/A
NP2	N/A	N/A	N/A	N/A	N/A
NP3	N/A	N/A	N/A	N/A	N/A
NP4	IEEE 1547	Power quality and safety	• AS 60038 • AS 61000.3.100 • AS 61000.2.2 • AS 3000 • AS 61000	Cyber security	NERC CIP, ISO/IEC 15408 (Common Criteria), ISO/IEC 27002:2005
NP5	N/A	N/A	N/A	N/A	N/A
NP6	N/A	N/A	N/A	N/A	N/A
NP7	N/A	N/A	N/A	N/A	N/A
NP8	N/A	N/A	N/A	N/A	N/A
NP9	N/A	N/A	N/A	N/A	N/A
NP10	N/A	N/A	N/A	N/A	N/A
NP11	N/A	N/A	N/A	N/A	N/A
NP12	N/A	N/A	N/A	N/A	N/A
NP13	N/A	Voltage Limits	• AS 60038 • IEC 60038	N/A	N/A
NP14	N/A	N/A	N/A	N/A	N/A
NP15	N/A	N/A	N/A	N/A	N/A
NP16	N/A	N/A	N/A	N/A	N/A
NP17	AS/NZS 4777	N/A	N/A	N/A	N/A
NP18	N/A	N/A	N/A	N/A	N/A
NP19	N/A	Installation and safety requirements (PV arrays)	AS 5033	N/A	N/A
NP20	N/A	N/A	N/A	N/A	N/A

Table A.7 Digital twin technologies and data acquisition systems in the national microgrid's projects

Code	Digital twin technology	Applications	Physics-based simulation	Physical system/type	Data type	Data acquisition system	Communication technologies
NP1	N/A	N/A	N/A	N/A	N/A	N/A	N/A
NP2	Adaptive GUI	• Design, monitoring, and • Operation management	N/A	N/A	N/A	Smart meters	Wireless
NP3	N/A	N/A	N/A	N/A	N/A	N/A	N/A
NP4	N/A	N/A	N/A	N/A	N/A	Smart meters	A fibre optic network
NP5	N/A	N/A	N/A	N/A	N/A	Smart meters	N/A
NP6	N/A	N/A	N/A	N/A	N/A	N/A	ComAP software for wired communication between diesel generator controller and ABB control system for solar power and battery management
NP7	N/A	N/A	N/A	N/A	N/A	SCADA and Cloud-prediction technology	N/A
NP8	N/A	N/A	N/A	N/A	N/A	NMI	N/A
NP9	Adaptive GUI	• Design, monitoring, and • Operation management	N/A	N/A	N/A	SCADA	N/A
NP10	N/A	N/A	N/A	N/A	N/A	N/A	N/A
NP11	Physics based simulation	Forecast the expected supply and demand	Operator training simulator (OTS) is used to simulate microgrids in different scenarios	N/A	N/A	Smart meter	3G/4G network
NP12	N/A	N/A	N/A	N/A	N/A	N/A	N/A

(continued)

Table A.7 (continued)

Code	Digital twin technology	Applications	Physics-based simulation	Physical system/type	Data type	Data acquisition system	Communication technologies
NP13	N/A	N/A	N/A	N/A	N/A	• High voltage: SCADA • Low voltage: AMI and Wattwatchers (WW) • Environmental parameters: Sky camera, pyranometer, weather station	N/A
NP14	N/A	N/A	N/A	N/A	N/A	Western Power's PI System	N/A
NP15	N/A	N/A	N/A	N/A	N/A	Smart meters	N/A
NP16	N/A	N/A	N/A	N/A	N/A	NMI	N/A
NP17	Physics based simulation	Forecast the average daily load profile by assuming the daily load and meteorological data	Response of PV, BESS and grid is analysed for forecasted average daily load profile	Interconnection of PV, BESS, grid, EV charging station and residential load	Power and state of charge	NMI	Wireless communication
NP18	N/A	N/A	N/A	N/A	N/A	N/A	N/A
NP19	N/A	N/A	N/A	N/A	N/A	N/A	N/A
NP20	N/A	N/A	N/A	N/A	N/A	N/A	N/A

Table A.8 Business models of the national microgrid's projects

Code	Business model—actor/type	Type of market	Ownership type	Tariff structure
NP1	N/A	N/A	N/A	N/A
NP2	Utility: design, build, own, operate, and maintain (DBOOM) Retailer: N/A	Retail market	Single owner (government ownership)	Time varying tariff
NP3	N/A	N/A	N/A	N/A
NP4	Long run marginal cost approach	Retail market under NREL	Multi-user local ownership	N/A
NP5	N/A	N/A	N/A	N/A
NP6	DER owner: selling renewable energy to the mine owner under 5.5-year power purchase agreement	N/A	Single owner (third party ownership)	N/A
NP7	N/A	N/A	Multi-owner (the solar plant is owned by IIG Solar Income Fund)	N/A
NP8	• Virtual microgrid • Participants: local distributed energy resources (PV or batteries) and demand response • Incorporate into a local marketplace	Peer-to-peer energy trading (blockchain-based software platform), and demand response	Multi-owner (virtual) microgrid	Time-varying and feed-in tariff
NP9	DER owner: design, build, own, operate, and maintain (DBOOM)—a significant step has been taken in WA to move towards hydrogen economy	N/A	Single owner	N/A
NP10	N/A	N/A	N/A	N/A
NP11	Private entity: design, build, own, operate, and maintain (DBOOM)—microgrid can sell renewable energy in the wholesale market and provide voltage and frequency support to the grid	Transactive energy markets	Single owner	Time varying optimised tariff
NP12	Utility: Design, Build, Own, Operate, Maintain (DBOOM)—supply power to military sector	N/A	Single owner (government ownership)	N/A
NP13	N/A	N/A	N/A	Feed-in tariff
NP14	N/A	N/A	N/A	N/A

(continued)

Table A.8 (continued)

Code	Business model—actor/type	Type of market	Ownership type	Tariff structure
NP15	• Peak demand saver plan: • Virtual energy storage + Time of use (ToU) pricing + solar sell back credits	Retail market	Multi-owner	• Time of Use (ToU) pricing: • Off-peak day (midnight–4 pm) and off-peak evening (8 pm–midnight): 26.8145 c/kWh • Peak daily (4 pm–8 pm): 51.1995 c/kWh
NP16	• DER owners (greater than 1 MW): sell energy and ancillary services to wholesale market • Community leadership organisation: advocacy, facilitation, funding attraction and community engagement realms	Spot market, Frequency Control and Ancillary Services (FCAS)	Multi-owner (community-owned) microgrid	Time-varying
NP17	• Retailer: rate-base • The retailer is allowed to sell energy to residents who use less than 50 MWh of energy per year in the wholesale energy market • Utility: DER owner	Wholesale Energy Market (WEM)	Multi-owner	• Supply charge (per meter per day) is $1.0155 • Electricity charge (per kWh) is $0.2833
NP18	Third party ownership/DER owner sell renewable energy to mine owner under a long-term power purchasing agreement	N/A	N/A	N/A
NP19	N/A	N/A	N/A	N/A
NP20	N/A	N/A	N/A	N/A

Table A.9 Highlights, limitations, and recommendations based on the national microgrid's projects

Code	Highlights/outcomes	Limitations	Recommendations
NP1	• Equipped with flywheel storage to stabilise power system to allow higher renewable penetration • Reduces diesel usage by adding a short-term spinning reserve • The levelised cost of energy for solar/diesel or wind/diesel systems is lower than just diesel • The challenge is standardising system designs and reducing integration costs	N/A	• It is important to size renewable energy generation capacity to maximise ROI and fuel savings • Future of power generation control—distributed control system • Higher penetration wind/diesel systems—"Smart Grid" technology
NP2	• It is the world leading project that provides reliable and secure electricity with high proportion of renewable energy (Up to 65%) • It provides 100% of power from renewable energy sources when conditions are suitable • On average scale in a year, ciesel generator is turned off for 20% of the time	• Carbon dioxide emission is reduced but not eliminated in this project • Excess renewable energy is not utilised in an efficient way, and it is delivered to the dynamic resistors that represent the artificial load for load balancing	• Up gradation of wind farm • Deployments of carbon block energy storage system to reduce dependence on diesel generators • Replacement of diesel fuel by liquefied natural gas • Deployment of ocean power system
NP3	• The system is more reliable, and renewables reduce the effects of rising oil prices • Horizon Power has been able to share its knowledge of how to design and build a renewable system with communities worldwide via Marble Bar and Nullagine, which have both been recognised as leaders in renewable integration • Local residents in these isolated communities better understand how much energy they use every day • Renewable sources produce 30% of the towns' annual energy demand • Reducing greenhouse gas emissions by 1100 tonnes per year	N/A	N/A

(continued)

Table A.9 (continued)

Code	Highlights/outcomes	Limitations	Recommendations
NP4	• Customers are generally very positive about the concept of edge-of-grid microgrid, the inherent sense of efficiency, and the innovation and associated it with a higher quality development • Price is the key motivator of acceptance and customers are attracted to the co-operative structure with dividends • Most consumers' concerns are due to mis-interpreting the packages • People prefer not to maintain their equipment	• Consumers lack-of understanding of microgrids • Appropriate regulatory framework for multi-user local ownership model for microgrids • Lack of clarity in legislation for private electrical infrastructure in public roads • Lack of common technical standards for microgrids • Lack of recognition of the social, economic and environmental benefits of microgrids in the process of allocation of government infrastructure funding arrangements • Dealing with the franchise rights of Ausgrid (the local distribution company) to operate as a monopoly in areas designated to them under NSW Energy Supply Act • The need for clear operator of last resort framework for independent distribution companies • Educating and working with the home builder supply chain rather than just the developer is essential for maximising the energy efficiency and demand management potential	To Establish: • A clear regulatory framework for microgrids • The rules of engagement for microgrid in terms of retail, generation and network operation and ownership • Clear rules of engagement between existing distributcrs and the new microgrids
NP5	• Getting the balance right • Storage can help the network • Energy sharing can help the balance • The national electricity market's overall electricity demand has been decreasing, and summer peak demand has increased significantly • For front-of-meter projects to be viable, they must reflect the benefit to the network operator as well as the consumers • Microgrids and virtual power plants provide network support by reducing the total load on the network, or by supplying local power generation or power quality services • Benefit related to 'no sharing' scenario is solar load shift • Benefits related to 'shared energy' scenario are solar load shift, tariff arbitrage, and wholesale price response • Benefits related to 'shared energy and demand management' scenario are: solar load shift, tariff arbitrage, wholesale price response, network response, and peak demand management • Costs related to 'shared energy' and 'shared energy and demand management' scenarios are upfront capital for solar and batteries, control systems and operating costs	• It is crucial for front-of-meter solutions to grasp the technology and use supply and demand principles to determine which areas offer the greatest value • Providing 100% renewable electricity requires better coordination of supply and demand across the grid • Maintaining the utility network is a challenge due to its size • In any network with reduced maintenance or redundancy, a major problem is not an arbitrarily noisy power supply, but large failures at high network usage times	N/A

(continued)

Table A.9 (continued)

Code	Highlights/outcomes	Limitations	Recommendations
NP6	• It assists mining businesses in assessing the hazards of high penetration renewable energy integration into diesel power facilities • It shows the potential diesel and cost reductions that renewable energy can provide in a mining environment • It analyses the cost curve and critical events that could allow high penetration solar PV potential in Australia to be economically installed without subsidies	• When an individual genset ComAp controller was disconnected, a bug in the ComAp Intelisys Bridge software caused erroneous communication between the ComAp and ABB control systems • Diesel generator requires necessary maintenance after specific time period	• To prevent erroneous communication between the ComAp and ABB control systems, bugs in the ComAp software must be removed • The remote monitoring workers may be trained to help with remote reaction to alerts and malfunctions • Personnel performing maintenance on relevant equipment must notify all stakeholders prior to the start of the operation • The financial impact of the project can be derived from its operational data that can be helpful for the pre-planning and design stage of the similar projects in future
NP7	• The battery storage system and cloud-detection system have provided a smooth ramp-up and ramp-down to PV generation output in order to minimise the energy curtailment and related financial impact • The cloud-detection events which requires maximum battery capacity are limited in time • The maximum battery output has been called nine times at most in one day	• The prediction accuracy is highly impacted by any dirt and dust on the cloud camera. The maintenance plan includes cleaning sensor lens twice a year • There were initial intermittent data issues with the controller. The main identified cause is the large number of data requests to this controller from the data logger. The solution could include reducing the number of parameters being logged • The 1-s time intervals over the huge amount of data produced by sensors should be limited in so that Excel could manage datasets effectively. Database solutions as well as changing time resolution are among the solutions • The battery system data and the data received from the main meters and PV plant are not collected on the same system. Although difficult and time-consuming, combining the data to get an overall view of the operation would be useful	• The real-world cloud prediction technology (CPT) systems is subject forecasting uncertainty, and the control system designer should consider the balance of both false positive and negative predictions and their associated risks • When the CPT predicts an occlusion event, the PV array is curtailed to a fixed set point. If the curtailment set point is too high, the storage system may not have enough power to compensate for an imminent drop in power, while the low curtailment set point may result in untenable amounts of lost yield • The impact upon ramp rates due to the charging power level, and the method of state of charge control deserve a considerable consideration when a renewable energy system is being designed
NP8	• The LEM could help consumers save 6–12% on energy purchases • Prosumers could make between 18 and 37% more money than selling surplus electricity to their retailer under the existing feed-in-tariff • LEM operates as a true market, not one necessarily controlled by a single retail intermediary, providing a market platform for peer-to-peer energy trading, demand response, and virtual power plants	• Identification of the location and estimated duration of a system limitation during the forward planning phase • An estimate of the costs associated with the proposed network solution for remedying the system limitation • The amount that the peak demand at the location of the system limitation would have to be reduced, as well as the dollar value that deferral would add to the network • Microgrid operation in islanded mode is not investigated • Surplus energy from the microgrid operator point of view and developing a distribution level market for buy/sell needed/surplus energy by microgrid • Utilising P2P infrastructure for running energy management schemes is not considered • Costs associated with implementing and administering the LEM platform is not considered	• Encourage the Australian Energy Market Commission (AEMC) to keep working on developing a system that would allow demand response to be a part of wholesale electricity markets • Assist AEMO in its investigation of the benefits and costs of allowing distributed energy resources (DERs) to access wholesale markets and ancillary services • Enable data sharing among participants while ensuring privacy and confidentiality • Ensure opt-out cost-reflective pricing for networks and remove structural barriers for cost-reflective pricing adoption

(continued)

Table A.9 (continued)

Code	Highlights/outcomes	Limitations	Recommendations
NP9	• This study demonstrated how a commercial-scale microgrid incorporating solar, battery, and gas production can overcome availability and variability challenges inherent in intermittent renewable energy technologies by working synergistically with other components within the microgrid • In this project, solar PV and a battery energy storage system (BESS) were combined with the site's existing natural gas-fired engine • A cluster controller, which is synchronised with the Western Power distribution network, has been used to regulate system balance • The system is intended to separate the microgrid from the central distribution network, thereby 'islanding' it	• From the consumer's standpoint, there is a lack of knowledge about microgrids • Lack of clarity in the correct legislative framework on the perception of microgrids with multi-user local ownership models • In the process of allocating government infrastructure financing arrangements, there is a lack of understanding of the social, economic, and environmental paybacks of microgrids	• The data will give technical insights into how hydrogen might be used as a future balancing fuel for use in the electricity system • A new purpose-built training centre will be used to test hydrogen applications and upskill people in readiness for the future energy workforce
NP10	• Design and installation both reflected the extreme conditions experienced at the site, including extreme heat, cyclonic conditions, and a corrosive environment • Installation and commissioning required minimal impact on the adjacent Onslow Power Station which was critical to maintaining a reliable power supply to the Onslow township • On completion the Onslow DER Project will be one of Australia's largest distributed energy resources (DER) microgrids with a very high level of renewable energy penetration • The solar farm was installed and commissioned with minimal impact on the adjacent Onslow power station which was critical to maintaining a reliable power supply to the Onslow township • The Onslow microgrid utilises a Distributed Energy Resource Management System (DERMS) to integrate traditional and distributed energy sources • The DER management system is set to utilise renewable energy to deliver up to half of the energy requirements	• Design considerations for extreme site conditions are challenging as the selection of suitable equipment with reliable performance in the long-term would be difficult • As Onslow is located in a region with severe tropical cyclones, the site must be shut down in response to extreme weather • Potential impact of extreme heat on exhaustion and fatigue especially during summer month should be minimised	Optimising the performance of customer batteries is being explored by issuing instructions in order for batteries to charge and discharge at the most appropriate periods during a day

(continued)

Table A.9 (continued)

Code	Highlights/outcomes	Limitations	Recommendations
NP11	• Indra's grid management platform will be used to control distributed energy resources in the Indra Monash Smart City project, which will include at least 1 MW of solar panels, 20 buildings, electric vehicle charging stations, and 1 MWh of energy storage • Indra's Active Grid Management (InGRID AGM) platform will provide real-time monitoring and control of grid-connected assets while being optimised to add value to customers, market participants, and the electricity grid • The Indra Monash Smart City will demonstrate the reliability of a 100% renewable powered embedded network, as well as the value it can provide to customers and the broader energy network	There is a lack of understanding of the social, economic, and environmental paybacks of microgrids in the process of allocating government infrastructure financing arrangements	A secure and private mobile network will be setup and connected to the cloud components of AGM in the near future of the TEM
NP12	• Operate independently or in cooperation with the Western Australian electrical network • Connect the microgrid to the Western Power network • The system's functioning also helps the grid power quality on HMAS stirling, which benefits defence and Western Power • Develop and create machine learning approaches to boost GIMG's performance	• Connection approval • Navigating and understanding systems and contractors during the project's design, construction, commissioning, and operation phases • During high load and generation events, manage the system's output to satisfy the base's requirements	• Upgrade the infrastructure of the existing network • Future wave energy integration

(continued)

Table A.9 (continued)

Code	Highlights/outcomes	Limitations	Recommendations
NP13	• The strong correlation between the overvoltage event occurrences and the PV generation profile is shown in this study • Plotting the number of high voltage events against the time-of-day of occurrence has a profile that correlates vigorously with the daily irradiance profile • The low voltage events take place in the afternoon or early evening during the residential peak load period • In spite of significant high voltage events, the presence of low voltage events indicates that simple tap changes to a lower voltage may solve one problem and create another • The study shows that highest aggregate PV output variations are mainly due to the widespread inverter tripping events, not cloud events	• Customer understanding of energy storage options. This is further complicated by the feed-in tariff • The industry needs to invest effort into installer education with respect to inverter settings. Changing network conditions due to increased penetration of inverter technology will see more frequent activation of primary response curves in inverters. This will change the nature of conversations between system retailers or installers and the prospective end customers • Installers should admit that the export of excess energy is an opportunity for customers, and they should not treat this ability as a right. There will be times when network conditions do not grant that opportunity, and their inverter will behave accordingly • As inverter primary response curves as front-line managers of DER, play an increasingly important role, ensuring the correct settings upon installation becomes critically important • Clear communication of the setting required by the network operator and easy access to that information by retailers and installers will play a vital role in ensuring the future fleet of next-generation inverters can play their part in preserving network stability	• In case high voltage events are related to one LV network and no low voltage events are observed for the HV feeder, lowering the LV transformer tap position would be a solution to reduce the voltage at the secondary side of the transformer • It is seen that the South Carnarvon HV feeder had fewer high voltage events compared with other HV feeders, but the highest number of low voltage events. In this case raising the tap position of the transformer is recommended • One of the HV feeders recorded the second-highest number of high voltage events from its WWs. Also, some low voltage events are observed. While another LV feeder (Gibson) had the most events. Since the LV feeder (Gibson) have both high and low voltage events, changing the transformer tap setting for LV networks would not be suggestion. In this situation using the voltage response function of PV inverters, which are supported just by inverters developed after 2015, is recommended. Other solutions could include the use of distribution transformers with the ability of online tap changing and line drop compensation capabilities • Gaining the level of visibility and control over distributed energy generation that we have had over centralised power stations, help us manage the contribution from DER to increase hosting capacity and achieve reliable and robust operation of microgrids

(continued)

Table A.9 (continued)

Code	Highlights/outcomes	Limitations	Recommendations
NP14	• Having a microgrid improves the reliability of power sources significantly • A microgrid can improve the Kalbarri power supply without additional generation • Western Power will gain a detailed understanding of how microgrid technology can be deployed cost-effectively to improve consumer outcomes by participating in the development of a microgrid and serving as a member of the management team • Analysing system stability, steady-state, and system protection	• This feasibility study did not include an analysis of the final optimised design and composition of a microgrid solution, and, therefore, a detailed forecast of its ultimate cost • The network is assumed to have no constraints	• Western power or 3rd party could purchase or hire a diesel generator to cover the maximum peak demand for the year or peak tourist times, respectively • System stability studies of microgrids for voltage and frequency stability for different configurations in both islanded and grid-connected modes • Voltage stability studies should include voltage recovery after faults, voltage recovery on motor starting, voltage step after generation/load trips, and black starting on battery • Frequency stability studies should include the contingency frequency control capability, stored energy, battery delays, and regulatory frequency control • A frequency sweep at the connection point between the battery and the generator is needed to observe resonance frequency • Developing a microgrid at Kalbarri requires Western Power to prepare a business case and undertake detailed planning and procurement processes
NP15	• A community battery is perceived as providing increased house value • New energy technologies and battery storage were the interests of the vast majority of respondents • Most of those in the trial were satisfied, and the main reason for this was lower energy bills • The top three benefits of Alkimos Beach's energy storage trial were lower bills, better utilisation of the rooftop solar system, and a greener future • The order of importance for signing up for the trial was: reducing bills, getting use out of solar PV system, the environment, curiosity, benefits of battery without the cost, independence from the grid, being invited by Synergy, being part of the community, and what others were doing • ABEST mostly benefits respondents in lowering energy bills without having to invest in an expensive home battery • Virtual battery products are best for customers who export during the day and consume during peak hours • Average participant saving per bill during the project lifetime is $35.83	• Energy markets are evolving, so at the time, community batteries did not have the frameworks in place to participate in every segment of the electricity market • The battery was set up with a 'set and forget' charge and discharge cycle, but it could have been dispatched more efficiently using smart controllers for smart dispatch	• A comprehensive survey is necessary to understand how residents have perceived the trial to shed light on both the current trial and possible future community battery initiatives • A realistic assessment of potential savings for different households with different usage profiles and ongoing support may be useful for empowering current customers, enticing new customers, and boosting the overall positive sentiment about the project • Collaborate regularly and transparently with stakeholders and counterparties

(continued)

Table A.9 (continued)

Code	Highlights/outcomes	Limitations	Recommendations
NP16	• Rooftop solar penetration was higher than the national average (57.6%) • Established three microgrid projects in the township (solar + battery, controllers, etc.) • Australia's first community-scale, grid-faced solar-battery system • Between 2004 and 2020, the average PV installation size increased significantly from 1.58 to 6.56 kWp • The Net Present Value (NPV) of all options in this study appears to be significantly negative. If financial/commercial goals were driving motivation, none of these options are likely to be pursued • Negative NPV indicates how much grant or philanthropic support a project requires to be financially viable • The LCOE calculations predict that all five options have relatively high life-cycle costs. In an emergency climate context, these options deserve consideration • Considering all of the options included in this study (and assuming current/projected demand/consumption rates), Yackandandah is unlikely to achieve a 100% gross target • With the current high level of solar penetration and the high growth rate of BTM solar generation, Yackandandah will reach 100% net renewable by 2027 • Yackandandah will be 100% net renewable by 2024–2025 under all five of the options examined	• As there are relatively small landholdings/high density land use in the Yackandandah Valley and highly productive agricultural land, finding suitable locations to install renewable energy equipment is difficult. Further, the community is concerned about the visual and aesthetic effects • Several other potentials and emerging technologies exist (e.g., hydrogen, compressed air, CO_2 storage) and have not been analysed • It was not included in this study's scope to evaluate demand-side and energy efficiency measures	• Incentives for BTM generation/storage, including a community subsidy scheme • It would be beneficial to investigate options and possible funding opportunities for grid-facing motorised generators to enhance emergency resilience

(continued)

Table A.9 (continued)

Code	Highlights/outcomes	Limitations	Recommendations
NP17	• Through innovative design, technology, and shared governance, this project laid the groundwork for increasing market uptake of solar photovoltaic (PV) energy solutions within strata developments in Australia • The project created a governance framework to overcome barriers that prevent solar PV panels from being installed in medium and high-density residential developments • The project also explored the challenges associated with electric vehicle charging and electricity sharing and trading approaches • This project has demonstrated the benefits of unbundled network costs for sharing DER on a local network from a policy standpoint	• There are some barriers in the whole sale energy market that limit some part of the implementation model • For the total amount of renewable generation required, there is less available solar PV installation area on the roof in high-rise apartment building • High-rise apartment buildings with PV would necessitate heavy equipment for installation, necessitating additional safety precautions and raising capital costs • Renewable energy installation and transmission become more expensive, and sometimes unaffordable. As a result, investors are hesitant to invest in renewables for these developments	• More work is needed in other states and territories to assist the rollout of the governance model across Australia, particularly in the areas of community titling and rulings on embedded network operators • Any future projects should include at least one complete billing cycle of hands-on training with strata managers (or whoever would be in charge of power billing/trading) during the commissioning phase • It is suggested that residents, strata managers, and any other interested parties involved in such shared solar-battery initiatives prepare a clear timeline of education and engagement activities, replete with critical milestones. Whenever possible, all hardware (PV systems, inverters, and batteries) should be purchased from an equipment list that has been pre-certified by Australian Standards, the network operator, and the applicable regulatory agencies. Where non-certified hardware is required, the programme should allow extra time for the certification procedure to be completed • To estimate the benefits of deploying behind-the-meter generation and storage on network infrastructure provision, more research is needed. Its impact on the 'Design After Diversity Maximum Demand' (DADMD), which is the maximum demand that the electrical distribution network (local transformer) can meet stated as a per-dwelling average
NP18	• Advanced Energy Resources (AER) was in charge of the facility's project concept, design, construction, and ongoing O&M • The project employs a novel back-to-back inverter topology that decouples load and renewable energy generation from the electricity network via a direct current link while allowing renewable energy export into the Western Power network	N/A	• As a result of AER's experience with the Port Gregory project, AER anticipates that many aspects of these future projects will be made easier and more cost effective • Following our learnings at Port Gregory, AER has also conducted significant design optimisation and anticipates a reduction of at least 25% in project capital costs as a result of optimised BESS design and implementation and a greater understanding of the wind turbine refurbishment and recommissioning process, particularly around logistics and technical recommissioning

(continued)

Table A.9 (continued)

Code	Highlights/outcomes	Limitations	Recommendations
NP19	• The output of generations are generally controlled based on an MGC mechanism via communication network • In case of communication loss, the storage system plays its role to adjust the system frequency, thus changing the generation level through pre-loaded power-frequency droop settings • The operation of diesel units is controlled by the Woodward easy generator controllers according to pre-set criteria • Diesel generators Start when the Battery State of Energy (SOE) reaches 5%, especially during the night when the storage system would be the primary source of energy; however, the storage system always remains grid forming, meaning that it provides the frequency sync signal for the generators • When the storage system reaches 90% SOE, the MGC automatically and gradually lessens excess solar generation as long as 95% SOE is obtained • HMI provides monitoring of PV inverter, the storage system and diesel generator performance in a real-time fashion	• System electrical protection: Inverter based generation produces low short circuit currents during fault conditions. As such protective systems normally provided on circuit breakers sized for nominal current and overload protection would not be able to safely protect by tripping, during three phase faults or single phase to ground faults • Fencing: The area of installation is situated next to a section of woodland, near to breeding area of flesh footed shearwaters, and the area must not contain any barbed wire to avoid entrapment of fauna (particularly ground dwelling birds), while according to AS 5033, the access to PV arrays should be restricted as their DC voltage may exceed 600 V	N/A
NP20	• Designing and constructing a containerised high-penetration hybrid energy system utilising wind and solar generation • Integrating supporting technologies for containerised energy hub and developing its technical capability and commercial viability for remote and island communities, thus reducing the operation and capital cost • Minimising the physical impacts on the existing power station and providing opportunities in order for the capital cost to be reduced up to 30% over a five-year period	• A supplementary protection equipment was required to strengthen protection schemes for fault detection with the aim of isolating the impact of possible outages. In this regard, physical measures, including reducing spans between poles and increasing horizontal separation, are considered to reduce number of faults caused by line clashing • One of the challenges caused by large renewable installations across the system is to economically capture the spill of energy • Due to the emerging battery technologies, it will be challenging to determine batteries lifespan according to the frequency adjustment cycles of the system, and their charging levels management	N/A

Appendix B

I. A. Ibrahim et al., *Microgrids Design and Operation*, Power Systems,
https://doi.org/10.1007/978-981-96-4914-3

Table B.1 Capability review of several international microgrid's projects

Code	Project	Year	City, state	Category	Type	Operation mode	Scope/objectives	Key use case(s)
IP1	Sifnos Island Cooperative [26, 27]	2016	Sifnos Island, Greece	Residential—(small island-tourism industry)	AC	Islanded	• By utilising DRESs, the project demonstrates an autonomous renewable system and sustainable future for the energy community • Identify the social and economic implications of the energy transition in isolated, insular communities • Achieving 100% energy independence • Cost reductions associated with electricity production • Removing the island's dependency on thermal generators and imported fossil fuels	• Balancing supply and demand on different timescales • Supply-side management • Long-term planning of microgrid infrastructure design and upgrading • Forecasting energy production and consumption on different timescales • State analysis and sub-sequent actions for providing operation in islanding mode
IP2	Remote area Energy supply with Multiple Options for integrated hydrogen-based technologies—DEMO 1: Ginostra [28, 29]	2018–2023	Ginostra, Italy	Residential (small village)	DC	Islanded	• Maximise the utilisation of local renewable energy resources • Reduce diesel consumption to lower local pollution and electricity costs • A more reliable power supply	• Matching demand and supply on different timescales • State analysis and subsequent actions for providing operation in islanding mode • Supply-side management • Storage management
IP3	Remote area Energy supply with Multiple Options for integrated hydrogen-based technologies—DEMO 2: Agkistro [28, 29]	2018–2023	Agkistro, Greece	An agri-food processing unit	DC	Islanded	• Making the building energy-autonomous and avoiding grid connection • Avoid an expensive grid connection investment (20 km away) • Avoid paying a high price for electricity from the grid • A more reliable power supply	• Matching demand and supply on different timescales • State analysis and subsequent actions for providing operation in Islanding mode • Supply-side management • Storage management
IP4	Remote area Energy supply with Multiple Options for integrated hydrogen-based technologies—DEMO 3: Ambornetti [28, 29]	2018–2023	Ambornetti, Italy	Residential (small village)	Hybrid	Islanded	• Building an off-grid community powered by RES with no dependence on fossil fuels • Avoid an expensive grid connection investment	• Matching demand and supply on different timescales • State analysis and subsequent actions for providing operation in Islanding mode • Supply-side management • Storage management

(continued)

Table B.1 (continued)

Code	Project	Year	City, state	Category	Type	Operation mode	Scope/objectives	Key use case(s)
IP5	Remote area Energy supply with Multiple Options for integrated hydrogen-based technologies—DEMO 4: Froan/Rye [28, 29]	2018–2023	Froan/Rye, Norway	Farms (remote islands)	AC/DC	Islanded	• Energy independence for the islands • Avoid replacing the outdated sea cable, which is costly and invasive • Avoid diesel generation due to its pollution issues and high costs (Froan is a natural reserve)	• Matching demand and supply on different timescales • State analysis and subsequent actions for providing operation in Islanding mode • Supply-side management • Storage management
IP6	Techno-Economic Feasibility Analysis of a Microgrid in Downtown Cleveland, Ohio [30]	2018	Downtown Cleveland, Ohio, USA	Commercial and industrial	AC	Grid connected and islanded mode	• To provide consumers an increase in electrical system uptime for up to 99.999%, or less than 6 min of power loss per year • To improve the power quality and reduced energy cost for backup power system	• Storage management • Microgrid provides islanding mode • Planning • Demand-side management
IP7	Demonstrating a secure, reliable, low-carbon community microgrid at the Blue Lake Rancheria [31]	2019	California, USA	Commercial	AC	Grid connected and islanded mode	• To design and develop a microgrid with reduced energy cost and greenhouse emissions • To provide robust backup power to critical loads during emergency situations • To transfer the knowledge gained from this project to broader audience • To obtain renewable energy production that exceeds 17% of annual on-site energy consumption • Over the course of one year, the objective is to save at least 25% on energy costs • To reduce annual greenhouse gas emissions by at least 195 tonnes of carbon dioxide equivalent (CO_2e) in the first year of operation	• Storage management • Microgrid provides islanding mode • Planning • Demand-side management • Protection and restoration • Enhance local power quality

(continued)

Table B.1 (continued)

Code	Project	Year	City, state	Category	Type	Operation mode	Scope/objectives	Key use case(s)
IP8	Borrego Springs: California's First Renewable Energy-Based Community Microgrid [32]	2019	California, USA	Residential, commercial, and industrial	AC	Grid connected and islanded mode	• To provide an islanding mode of operation during daytime and to reduce the non-critical load during night, • To provide a customer an increase in uptime service of electricity up to 98% at lowest energy cost • To provide renewable energy to 100% of the community load • To reduce the carbon dioxide emissions by more than 20% as compared to the emission from fossil fuel based distributed generator set	• Storage management • Microgrid provides islanding mode • Planning • Demand-side management • Protection and restoration • Enhance local power quality • Microgrid provides black start feature • Microgrid provides time to live feature of a resource • Microgrid provides alarm feature for any monitored value that go beyond the acceptable limits
IP9	A Novel, Renewable Energy Microgrid for a California Healthcare Facility [33]	2019	California, USA	Commercial (A non-for-profit hospital)	AC	• Grid-connected and islanded (In grid-connected mode around 7% of load are supplied by PV generation while 93% supplied by the main grid) • Islanding mode capability is limited to emergency	• Identify implementation challenges towards renewable energy microgrid for healthcare facilities • Reduce hospital energy consumption and fossil fuel usage, especially during peak load period, and minimise the costs • Provide islanding operation mode during utility power outages and emergency conditions • Develop and accomplish an innovative control system for the microgrid which is able to be commercialised • Propose a novel dual interconnection methodology	• Microgrid provides islanding mode • Balancing supply and demand on different timescales • Storage management • Supply-side management
IP10	Las Positas College Microgrid [34]	2019	California, USA	Educational and commercial	AC	Grid-connected	• Develop applications and evaluate performance of implementing smart grid systems and processes • Enhance the capabilities of microgrid energy management using storage systems and other equipment • Use a novel battery energy storage system with controllers to reduce peak demand and manage energy use • Provide information and data on innovative technologies, especially controllers and storage systems for educational institutions	• Microgrid Operator sells/buys energy on external market • Microgrid provides islanding mode • Balancing supply and demand on different timescales • Storage Management • Matching demand and supply on different timescales • Automated Load Shaping • Continuous determination of available controlling power range on different timescales

Table B.2 International capability review of microgrid's technologies, modelling techniques and tools

Code	Network topology	Network data	Technology	Technology data	Time horizon	Modelling technique	Modelling tool
IP1	Radial/remote island	N/A	Wind turbine	Wind velocity	Hourly average values	Weibull distribution	N/A
			Pumped hydro storage	Use of seawater	N/A	N/A	N/A
			Load	• Normal loads • Charging load of future EVs (controllable load)	Hourly average values	• Power demand annual time series • Aggregation of the current power demand and the future EVs	N/A
			Hydrogen storage	Fuelling the maritime connection and producing potable water with hydrogen from the hybrid power plant's surplus electricity	N/A	• Annual power surplus time series is introduced • Alternate scenarios are implemented iteratively for the electrolysis unit's nominal power and the storage capacity of the storage tanks	N/A
IP2	Radial/remote island	N/A	PV power plant (170 kW)	Connected to DC bus	N/A	N/A	N/A
			Hydrogen storage	• DC/DC converter (efficiency 95–98%) • Fuel cell and electrolyser—units of 25 kW • P2G—Alkaline—50 kW • G2P—PEM—50 kW • Hydrogen storage—total gross energy: 1793 kWh	N/A	N/A	N/A

(continued)

Table B.2 (continued)

Code	Network topology	Network data	Technology	Technology data	Time horizon	Modelling technique	Modelling tool
			Battery storage (600 kWh)	• Energy buffer—alleviate the high-frequency variability of the RES • DC/DC converter (efficiency 95–98%)	N/A	N/A	N/A
			Load	• DC (bus)/AC (load) inverter (efficiency 94–97%) • Total annual load: 171.54 MWh	N/A	N/A	N/A
IP3	Backup network for the agri-food processing unit	N/A	Hydroelectric power plant (0.9 MW)	Connected to DC bus	N/A	N/A	N/A
			Hydrogen storage	• DC/DC converter (efficiency 95–98%) • Fuel cell and electrolyser—units of 25 kW • P2G—Alkaline—25 kW • G2P—PEM—50 kW • Hydrogen storage—total gross energy: 996 kWh	N/A	N/A	N/A
			Battery storage (30 kWh)	• Supporting the system operation (system start-up and operation of the control unit and auxiliary equipment) • DC/DC converter (efficiency 95–98%)	N/A	N/A	N/A
			Load	• DC (bus)/AC (load) inverter (efficiency 94–97%) • Total annual load: 87.4 MWh	N/A	N/A	N/A

(continued)

Table B.2 (continued)

Code	Network topology	Network data	Technology	Technology data	Time horizon	Modelling technique	Modelling tool
IP4	Radial/remote island	N/A	PV plant (40 kW)	Connected to DC bus	N/A	N/A	N/A
			Biomass-based combined heat and power (CHP) generator (50 kW)	Connected to DC bus	N/A	N/A	N/A
			Hydrogen storage	• DC/DC converter (efficiency 95–98%) • Fuel cell and electrolyser—units of 25 kW • P2G—Alkaline—25 kW • G2P—PEM—50 kW • Hydrogen storage—total gross energy: 498 kWh	N/A	N/A	N/A
			Battery storage (30 kWh)	• Supporting the system operation (system start-up and operation of the control unit and auxiliary equipment) • DC/DC converter (efficiency 95–98%)	N/A	N/A	N/A
			Load	• DC (bus)/AC (load) inverter (efficiency 94–97%) • Total annual load: 96.63 MWh	N/A	N/A	N/A

(continued)

Table B.2 (continued)

Code	Network topology	Network data	Technology	Technology data	Time horizon	Modelling technique	Modelling tool
IP5	Radial/remote island	N/A	PV plant (85 kW)	Connected to AC bus	N/A	N/A	N/A
			Wind turbine (225 kW)	Connected to AC bus	N/A	N/A	N/A
			Hydrogen storage	• Fuel cell connected to DC bus by DC/DC converter • Electrolyser connected to AC bus • Fuel cell and electrolyser—units of 25 kW • P2G—PEM—55 kW • G2P—PEM—100 kW • Hydrogen storage—useful gross energy: 3333 kWh	N/A	N/A	N/A
			Battery storage (550 kWh Li-ion)	• Energy buffer—alleviate the high-frequency variability of the RES • Connected to DC bus • DC/DC converter	N/A	N/A	N/A
			Load	• Connected to AC bus • Total annual load: 126.8 MWh	N/A	N/A	N/A
			Bus coupler	AC/DC converter to connect AC and DC buses	N/A	N/A	N/A
IP6	N/A	Block diagram of microgrid network containing local energy assets, vital community services and utility connection	PV	N/A	N/A	N/A	N/A
			Wind turbine	N/A	N/A	N/A	N/A
			Electrical battery	N/A	N/A	N/A	N/A
			Normal consumer load	N/A	N/A	N/A	N/A

(continued)

Table B.2 (continued)

Code	Network topology	Network data	Technology	Technology data	Time horizon	Modelling technique	Modelling tool
IP7	Rural	Simplified single line diagram of microgrid networks	PV (420 kW)	Power output per string per change in solar irradiance	Hourly and monthly	N/A	PSS SINCAL
			Diesel generator (1 MW)	N/A	5–15 min time interval	N/A	
			Electrical battery (500 kW/950 kWh)	Battery charge/discharge efficiency graph with respect to energy transfer	Hourly and daily	Linear regression is used to calculate the self-discharge with variable capacity of the battery	
			Controllable load (>300 kW)	Duration curve of controllable load (Gross and net side load)	5–15 min time interval	• Five types of controllable load on: • Office load (30 kW) • HVAC compressor load (30 kW) • Gaming load (100 kW) • Exhaust and HVAC unit fans (75 kW)	
IP8	Rural	Block and control diagram of microgrid network Microgrid architecture overview	PV (32.5 MW)	Solar irradiation data	Hourly	N/A	RSCAD
			Distributed generator (1.8 MW)	N/A	Hourly	N/A	
			• Electrical battery • (Three lithium-ion batteries of 0.5 MW, 1 MW and 0.25 MW respectively)	Battery charge/discharge efficiency graph with respect to energy transfer	Hourly	N/A	
			• Three types of loads: • Residential • Commercial • Industrial	Historical load profile	15-min time interval	N/A	

(continued)

Table B.2 (continued)

Code	Network topology	Network data	Technology	Technology data	Time horizon	Modelling technique	Modelling tool
IP9	Dual interconnection	• Operating voltage (480 V) • Single line diagram • PMU specifications	Solar PV (250 kW)	• Solar generation curve • Average daily PV production (862.93) • Peak PV production (1964 kWh) • Total PV production per month (25887)	1 Hz/60 Hz	N/A	• OSIsoft® Pi® platform • MATLAB® Simulink™
			Wind turbine	N/A	1 Hz/60 Hz	N/A	
			CHP	N/A	1 Hz/60 Hz	N/A	
			Battery system (250 kW/ 1 MWh)	N/A	1 Hz/60 Hz	N/A	
			Fuel cell (400 kW)	N/A	1 Hz/60 Hz	N/A	
			Diesel generators	N/A	1 Hz/60 Hz	N/A	
			EV charging station	N/A	15-min intervals	N/A	
			Heavy loads HVAC and ventilation systems	• Average daily energy consumption (12,330 kWh) • Average peak load (513.78) • Peak load baseline (750 kW)	15-min intervals	N/A	

(continued)

Table B.2 (continued)

Code	Network topology	Network data	Technology	Technology data	Time horizon	Modelling technique	Modelling tool
IP10	N/A	• System layout • Single line diagram • Operating voltage: (480 V AC) • PCC voltage (21.7 kV) • Switchboard (400 A) • Main breaker (2400 A) • Net power demand profile	Solar PV (2.3 MW)	N/A	15-min intervals	N/A	GELI EOS GELI-ESYST GELI DCM
			Thermal ice storage system (3,200 ton/h)	N/A	5-min intervals	N/A	
			Battery energy storage system (2x 100 kW/ 500 kWh)	N/A	Various: 1/5/15 min(s)	N/A	
			Loads: • Lighting (Light-emitting diode lighting (100 kilowatts)) • Heating • Ventilation • Air-conditioning	N/A	Various: 1/5/15 min(s)	N/A	

Table B.3 Power flow/state estimation techniques, analysis horizon and tools of the international microgrid's projects

Code	Power flow/state estimation	Balanced/unbalanced	Steady-state/dynamic	Snapshot/time-series	Steady-state PF tool	Dynamic PF tool
IP1	N/A	N/A	N/A	N/A	N/A	N/A
IP2	N/A	N/A	N/A	N/A	N/A	N/A
IP3	N/A	N/A	N/A	N/A	N/A	N/A
IP4	N/A	N/A	N/A	N/A	N/A	N/A
IP5	N/A	N/A	N/A	N/A	N/A	N/A
IP6	N/A	N/A	N/A	N/A	N/A	N/A
IP7	Power flow	Balance	Static	Time-series	PSS SINCAL	N/A
IP8	Power flow	Balance	Static and dynamic	Time-series	RSCAD	RSCAD
IP9	Power flow	Balanced	Static	Time-series	N/A	N/A
IP10	Power flow	Balanced	Static	Time-series	N/A	N/A

Table B.4 Optimisation problem definition, functions, and constraints of the international microgrid's projects

Code	Single- or multi-objective/ type	Optimisation data	Objective function(s)	Technical constraint(s)	Values	Economic constraint(s)	Values
IP1	Single-objective/ techno-economic	• Power demand annual time series for current and introducing consumption • Estimated wind generation • Pumped hydro storage characteristics	Optimal hybrid power plant sizing to maximise electricity demand coverage in real-time and on an annual basis (100%): • Wind park: 4× 3 MW • Hydroelectric power plant: 4× 2.185 MW • Pump station: 12× 0.857 MW	Wind power penetration (PRESp)	• If PRES > 40%.Pd, then PRESp = 40%.Pd • If PRES ≤40%.Pd, then PRESp = PRES	N/A	N/A
IP2	N/A	N/A	N/A	N/A	N/A	N/A	N/A
IP3	N/A	N/A	N/A	N/A	N/A	N/A	N/A
IP4	N/A	N/A	N/A	N/A	N/A	N/A	N/A
IP5	N/A	N/A	N/A	N/A	N/A	N/A	N/A
IP6	Multi-objective/ techno-economic	Sensitivity analysis data of microgrid net per value with respect to customer energy cost/MWh	• Minimising the cost of energy/kWh for 99.999% uptime time service • Maximising the power quality by reducing the harmonics, sag and swell in the voltage	N/A	N/A	The energy cost of 99.999% uptime service increases with increase in annual energy demand of the customers	Annual energy demand limit = 1 MW for each customer

(continued)

Table B.4 (continued)

Code	Single- or multi-objective/ type	Optimisation data	Objective function(s)	Technical constraint(s)	Values	Economic constraint(s)	Values
IP7	Multi-objective/ techno-economic	• Estimated monthly electricity bill without microgrid • Cumulative monthly microgrid saving data • Comparison graph of diesel power generation in each scenario verses PV and battery energy storage system	• Minimising the carbon dioxide emission • Maximising the energy cost savings • Minimising the energy consumption from utility grid	• The maximum available PV power at standard testing condition is higher than the maximum inverter output, this limit or clip the power production from PV system • Maximum export power limit to the utility grid as per interconnection agreement • In islanded mode, there is a limit on frequency and voltage deviation from the nominal values	• Maximum PV power at standard testing condition = 457.7 kW • Maximum Inverter output = 420 kW • Maximum export power = 100 kW • Voltage deviation in islanding mode = 5% from the nominal value • Frequency deviation from the nominal = 0.1% of the nominal value	The import limit for peak shaving	N/A
IP8	Multi objective/ techno-economic	• Monthly load profile data • Data of yearly sustained and momentary outages with its duration	• Maximising the uptime service of electricity up to 98% at lowest cost • Minimise the carbon dioxide emission by more than 20% as compared with the DG set • Maximising the power quality	• Voltage limit • Frequency limit • Voltage phase angle difference	• 3% of nominal value • ±0.1 Hz • ±10°	N/A	N/A

(continued)

Table B.4 (continued)

Code	Single- or multi-objective/type	Optimisation data	Objective function(s)	Technical constraint(s)	Values	Economic constraint(s)	Values
IP9	Multi-objectives/techno-economic	• Total usage • Peak demand • PV output power • Battery SOC • Utility costs • Power factor • Real/reactive power	• Maximising PV generation output • Minimising the cost of electric energy and demand charges • Financial optimising of battery state of charge	• Voltage limits • Physical space constraints	N/A	N/A	N/A
IP10	Single-objective/techno-economic	• PVs output power • Battery SOC • Price signals • Local consumption	Optimising time of energy purchase	N/A	N/A	N/A	N/A

Table B.5 Control and EMS technologies and tools applied in the international microgrid's projects

Code	Control level(s)	Local controllers/ technology	Secondary and tertiary control	EMS type (day-ahead/ online)	Monitored values	Operation optimisation function	Operation optimisation tool	Service provider
IP1	Primary and secondary	Wind turbine, and pumped hydro storage	Centralised	N/A	N/A	Maximising electricity demand coverage in real-time by wind and pumped hydro storage	N/A	N/A
IP2	Primary and secondary	Battery storage, fuel cell, electrolyser, PV solar, and diesel generator	Centralised	N/A	• Battery discharging and charging power • SoC of battery • Fuel cell power • Electrolyser power • H_2 storage SOC • Power provided by the external source (grid or engine) • Load consumption • RES generation	Maximising the benefits of the power-to-power (P2P) system by reducing the energy required from an external source (e.g., electrical grid or fossil fuel-based generators) • Reliable coverage of the loads • Protection of the different components and avoiding their operation outside safe limits	N/A	N/A
IP3	N/A	N/A	N/A	N/A	N/A	N/A	N/A	N/A

(continued)

Table B.5 (continued)

Code	Control level(s)	Local controllers/ technology	Secondary and tertiary control	EMS type (day-ahead/ online)	Monitored values	Operation optimisation function	Operation optimisation tool	Service provider
IP4	Primary and secondary	Fuel cell, electrolyser, PV solar, diesel generator, and CHP generator	Centralised	N/A	• Fuel cell power • Electrolyser power • H_2 storage SOC • Power provided by an external source (grid or engine) • Load consumption • RES generation	Maximising the benefits of P2P system by reducing the energy required from an external source (electrical grid or fossil fuel-based generators) • Reliable coverage of the loads • Protection of the different components and avoiding their operation outside safe limits	N/A	N/A
IP5	Primary and secondary	Fuel cell, electrolyser, PV solar, wind, diesel generator, and battery storage	Centralised	N/A	• Battery discharging and charging power • SoC of battery • Fuel cell power • Electrolyser power • H_2 storage SOC • Power provided by an external source (grid or engine) • Load consumption • RES generation	• Maximising the benefits of P2P system by reducing the energy required from an external source (electrical grid or fossil fuel-based generators) • Reliable coverage of the loads • Protection of the different components and avoiding their operation outside safe limits	N/A	N/A
IP6	N/A	N/A	N/A	N/A	N/A	N/A	N/A	N/A

(continued)

Table B.5 (continued)

Code	Control level(s)	Local controllers/ technology	Secondary and tertiary control	EMS type (day-ahead/ online)	Monitored values	Operation optimisation function	Operation optimisation tool	Service provider
IP7	Primary and secondary	Siemens Spectrum Power™ Microgrid Management System software—managing the management and control operation of microgrid in grid connected and islanding mode	Centralised	• Online	• Reactive power • Real power • Apparent power • Voltage • Frequency • Total energy • SOC of battery • Temperature • Humidity • Irradiation	Every 15 min, a 24-h optimal generation dispatch schedule is generated to minimise costs based on the electric rate schedule, short-term load forecast, and solar PV output forecast	Siemens Spectrum Power™ Microgrid Management System software	Environmental forecasting and monitoring server for real time weather data, weather forecast and solar forecast data
IP8	Primary and secondary	Advanced microgrid controller (Managing the management and control operation of microgrid in grid connected and islanding mode)	Distributed	Online	• Voltage • Frequency • Real Reactive power • SOC	Economic dispatch import and export limit on power (Peak shaving)	Advanced microgrid controller	Weather server for real time weather and weather forecast data

(continued)

Table B.5 (continued)

Code	Control level(s)	Local controllers/ technology	Secondary and tertiary control	EMS type (day-ahead/ online)	Monitored values	Operation optimisation function	Operation optimisation tool	Service provider
IP9	Primary and secondary	Internal inverter control provided by the Princeton Power Systems	Centralised	Online/ real-time	• Peak demand • Power factor • SOC of batteries • PV generation output • Active and Reactive power • Power quality	• Tunning internal inverter (primary) to maximise the PV output • Tracking the SOC reference profile by external controller (Secondary) for charging/discharging batteries considering the demand/supply balance • Machine decision-making	• OSIsoft® Pi® platform • MATLAB® Simulink™ • Windows Server 2012 (using Microsoft Visual Studio for C++ codes)	• Automated functions (energy arbitrage, photovoltaic power quality regulation, time-shifting, and peak load reduction) • A visualisation tool (GUI) for stakeholders to be able to track the system performance
IP10	Primary and secondary	• Energy storage systems • Loads	Centralised	Real-time Predictive Online	• Wind speed • Temperature • Active and reactive power • Voltage • Current • Frequency • SOC	To reduce energy use during demand response events to mitigate the utility peak demand Prevent islanding mode of operation by manual/ automatic switching when a short-term utility outage takes place Integrate the solar PV array into the islanded system as a second step	• GELI DCM and EOS Applications • Schweitzer Engineering Laboratories Axion Real-Time Automation Controller (RTAC)	• HMI for RTAC • A web-based solar and storage sizing and pricing application

Table B.6 Standards and guidelines for designing and operating microgrids in the international microgrid's projects

Code	Microgrid specific standards	Power grids		Supporting technologies	
		Category	Standard	Technology	Standard
IP1	N/A	N/A	N/A	N/A	N/A
IP2	N/A	N/A	N/A	N/A	N/A
IP3	N/A	N/A	N/A	N/A	N/A
IP4	N/A	N/A	N/A	N/A	N/A
IP5	N/A	N/A	N/A	N/A	N/A
IP6	N/A	N/A	N/A	N/A	N/A
IP7	IEEE 1547-2003	N/A	N/A	N/A	N/A
IP8	• IEEE 1547.1-2005 • IEEE 1547.8 • IEEE 1547, Section 4.4.1 • IEEE 1547-2003 • IEEE P2030.8	N/A	N/A	N/A	N/A
IP9	N/A	N/A	N/A	PMU communication	IEEE C37.118
				Inverter communication	Modbus TCP/IP protocol
IP10	N/A	N/A	N/A	N/A	N/A

Table B.7 Digital twin technologies and data acquisition systems in the international microgrid's projects

Code	Digital twin technology	Applications	Physics-based simulation	Physical system/type	Data type	Data acquisition system	Communication technologies
IP1	N/A	N/A	N/A	N/A	N/A	N/A	N/A
IP2	N/A	N/A	N/A	N/A	N/A	N/A	N/A
IP3	N/A	N/A	N/A	N/A	N/A	N/A	N/A
IP4	N/A	N/A	N/A	N/A	N/A	N/A	N/A
IP5	N/A	N/A	N/A	N/A	N/A	N/A	N/A
IP6	N/A	N/A	N/A	N/A	N/A	N/A	N/A
IP7	Physics based simulation	• Fault diagnosis in islanded mode • Short term load forecast • Solar PV output forecast	• Response of microgrid voltage during seamless connection and interconnection of diesel generator • Power and frequency response of microgrid during islanded mode	Microgrid network containing DER's and controllable load	Short circuit current	• SCADA • Electric power meters • GPS clock • PV data logger	Fiber-optic communication
IP8	Physics based simulation	Fault diagnosis	Hardware in the loop (HIL) emulation is used to perform physics-based simulations in different scenarios	N/A	N/A	• SCADA • Net energy meter • Power meter	Fiber-optic communication
IP9	• A C37.118 read interface run under MATLAB Simulink • A Modbus master/slave interface run under MATLAB Simulink	Real-time control	Hardware in the loop simulation	Microgrid (inverter) controller	• Experimental data set by PMUs • Power reference signals	• 6× PMUs 60 Hz At critical monitoring points • Meters for real power consumption • EMOS 1 Hz for inverter-connected devices	Internet
IP10	N/A	N/A	N/A	N/A	N/A	• Meters • Sensors	A VPN (virtual private network) for external internet communications

Table B.8 Business models of the international microgrid's projects

Code	Business model—actor/type	Type of market	Ownership type	Tariff structure
IP1	N/A	N/A	N/A	N/A
IP2	N/A	N/A	N/A	N/A
IP3	N/A	N/A	N/A	N/A
IP4	N/A	N/A	N/A	N/A
IP5	N/A	N/A	N/A	N/A
IP6	• Public-private partnership: • Utility is the DER owner and own the distribution system infrastructure • Operator is the private entity that own the control system	Spot power is purchased at location marginal pricing	Multi-owner (Utility and private entity)	• Small commercial customers (an average demand of 20 kW with less than 30 kW peak demand) = $140/MWh • Large commercial (an average demand of 100 kW with more than peak demand of 30 kW) = $120/MWh • Industrial customers (an average demand of 5 MW) = $90/MWh
IP7	Community is the owner and operator of DER's and distribution infrastructure	N/A	Multi-owner (Community)	N/A
IP8	• Utility is responsible for management and operation of microgrid • Distribution infrastructure is provided by the utility • Third party entity owns the microgrid asset	N/A	Multi-owner	N/A
IP9	Public-private partnerships (Charge Bliss has designed, engineered, built, and is operating the microgrid located at the Kaiser Permanente facility for a non-for-profit hospital)	N/A	Multi-owner microgrid	• Time-of-use energy Pricing: • 0.15245 $/kWh (12 pm–6 pm (peak)) • 0.08512 $/kWh (9:30 pm–8:30 am and weekends) • 0.11194 $/kWh (other periods) • Demand pricing
IP10	Private	• Wholesale energy market • Spinning reserves market • Demand response markets	Single owner	ToU

Table B.9 Highlights, limitations, and recommendations based on the international microgrid's projects

Code	Highlights/outcomes	Limitations	Recommendations
IP1	• Sifnos' levelised electricity production costs have been very high Because of the consumption of expensive diesel oil and the need to maintain spinning reserves, which required the diesel generators to operate with partial loads, reducing efficiency • The project is the first hybrid power plant at the global level to combine and operate a wind park and a Pumped Hydro Storage (PHS) system • Wind park and upper reservoir are located on the same installation site, which minimises the project's set-up costs and eliminates impacts on the natural environment • Using the proposed sizing and synthesis, the hybrid power plant achieves its fundamental objective of meeting 100% of the annual and real-time electricity demand • The PHS system is sized to offer an energy storage capacity of 860 MWh, implying a 16-day autonomy operation period based on daily average electricity consumption • Approximately 51.8% of wind park initial electricity production is surplus, which cannot cover power demand or be stored in the PHS system since the upper reservoir is fully charged. The island will benefit from it through its development activities • The project has minimal impacts on existing human activities and the natural environment due to the location of installation • Fuelling the maritime connection and producing potable water with hydrogen from the hybrid power plant's surplus electricity	The island's marine connection to the neighbouring islands is one of Sifnos' significant problems	• Maintaining a guarantee of power production on an island is the only way to ensure a secure energy supply on an island, regardless of whether it's interconnected or not • The supply of guaranteed power can be assured by either keeping the thermal generator(s) in standby mode or installing electricity storage units

(continued)

Table B.9 (continued)

Code	Highlights/outcomes	Limitations	Recommendations
IP2	• Hysteresis is also introduced to protect the battery from excessive use, thereby extending its lifetime, at the expense of more intensive exploitation of the electrolyser and fuel cell • Hybrid P2P reduces the current use of diesel generators by around 4.4% • Power is primarily supplied by the battery when RES power is insufficient to meet the load • Summer months see the SOC for hydrogen decline dramatically as almost all the stored gross energy in the hydrogen tank is used up • Energy cost reduction for end users	In the spring season, a considerable amount of renewable energy is required to be curtailed	• Implementing technical learning in product improvement • Creating new business by marketing and communicating results and experiences • A replication in other off-grid and isolated microgrids where competitive technologies (diesel generators) are not feasible economically or environmentally • The battery degradation phenomenon needs to be taken into account
IP3	In the case of Agkistro, the P2P solution acts as a backup system, providing energy autonomy for 1–2 days	N/A	N/A
IP4	• Since solar and biomass energy are sufficient for making the community energy-autonomous, outside power sources are not required • Hydrogen storage effectively maximises the RES exploitation since there is almost no curtailment • A reduction in CO_2 production of 57 tonnes per site per year • Create confidence with end users and authorities of isolated regions regarding the technology, business models, and market readiness	N/A	N/A
IP5	• Hysteresis is also introduced to protect the battery from excessive use, thereby extending its lifetime, at the expense of more intensive exploitation of the electrolyser and fuel cell • With local renewable energy and hydrogen/battery energy storage, energy needs from external sources can be reduced to below 5% of the annual load demand	A considerable amount of renewable energy must be curtailed in different seasons	N/A

(continued)

Table B.9 (continued)

Code	Highlights/outcomes	Limitations	Recommendations
IP6	• On the basis of indirect community value, such as emergency power, clean energy, and economic development, the proposed microgrid has been determined to be cost-effective • A microgrid that provides 99.999% uptime service while maintaining a 5% return on capital investment and a 3% return on investment could be built for around $0.13/kWh	• The occupied business model is not tax efficient model • The microgrid tariff structure is highly sensitive to the number of consumers and average demand	• Even though such indirect value is unlikely to be monetised, it is recommended to design a cost-effective model that will attract investors • It is recommended to examine the proposed cost model with the companion reports on market penetration, economic development, and the value of resiliency from the study group
IP7	• In 2017, this project saved $160,000 in electricity costs, a 25% reduction; Savings of $190,000 are expected in 2018 • This project has reduced greenhouse gas emissions by approximately 158 metric tonnes CO_2e, with an additional 170 metric tonnes CO_2e expected in 2018 due to operational improvements • It strengthened the resilience of a well-known Red Cross evacuation centre and demonstrated its ability to remain on an island for extended periods • There was a 10% increase in tribal employment as well as induced and indirect economic benefits • During numerous grid outages, it provided reliable, unattended backup power	• During low inertia islanding condition, a seamless connection of diesel generator with microgrid is only possible during testing due to some limitations in islanded mode of operation • Due to DC system limitation, the battery energy storage system has a maximum limit of 500 kVA output, it will immediately shut of in case of violation in maximum output limit	• In the project development stage, hire and train a project team with different technical skills • As early as possible in the project development process, maintain a collaborative liaison with the utility company • It is recommended to include advanced protection scheme to protect the microgrid at point of common coupling (PCC)

(continued)

Table B.9 (continued)

Code	Highlights/outcomes	Limitations	Recommendations
IP8	• Development of a sophisticated controller capable of isolating both the Borrego Springs community and all of their customers • The microgrid was managed locally and from utility control centre, which was more than 80 miles away • The proposed microgrid is successfully tested in an islanded condition that provide power to the whole community during daytime and night-time	• There are a lot of difficulties and challenges in controlling the microgrid assets that are owned by third part entity • The electric system becomes unstable in low inertia islanding condition	• Whenever possible, all possible testing should be performed at the test facility to simulate any potential problem encountered in field operation • Active resynchronisation capability is recommended to be included in the AMC in order to reduce magnitude, frequency and phase angle errors, the controller must adapt the voltage and frequency setpoints of the equipment acting as voltage and frequency master • RSCAD model for this project didn't contain the logic control function for capacitors. The analysation of controller dispatch capability becomes very difficult. It is recommended to include capacitor logic controller in RSCAD model • The regulation capability of the reactive power control can be improved by including the coordination of capacitor controller and voltage regulator within the microgrid controller

(continued)

Table B.9 (continued)

Code	Highlights/outcomes	Limitations	Recommendations
IP9	• Located in California, this project is the first renewable energy microgrid supplying healthcare facilities and showing the safe and reliable performance in this application • This project initiates and tests the interconnection of the life safety branch aimed at resiliency improving as well as at reducing the dependency on diesel generation. This shows that diesel generators are not the only solution to back up a system that supplies critical loads • A new autonomous control system has been developed using a novel monitoring strategy and control tools which can guarantee the safe, reliable and optimised operation of the microgrid systems • Performing demand management in peak hours, this project has been successful to enhance late day ramp rates which eliminates the need for a redundant generation	• The engineers have designed a new block wall room containing air-conditioning and internet communication systems, to accommodate some major parts of the microgrid (inverters, batteries) and control system as well. The reason lies behind the fact that there was no suitable place for a containerised system • The PV arrays are designed to be on top of the parking structure. This imposes an additional structure to the canopy that should resist the weight of the panels in different conditions • Owing to some regulatory restrictions, the microgrid is not able to export (sell) energy to the main grid. The reason lies behind the low ratio of system size to the base load of the hospital • Considering the low response time and system inertia, the microgrid can barely play a role in power quality regulation	• Given the importance of healthcare facilities, the development of back-up resources to support emergency circumstances is always of concern. This requires the standards consideration by the corresponding organisations. In addition, reducing reliance on fossil fuels is essential in providing back-up generation to increase the resiliency of healthcare facilities • Utilising multi-sources of energy in healthcare microgrids is another topic worth consideration in future research. These resources could be a renewable generation, including wind, solar, geothermal etc. or a continuous system such as fuel cell and CHP • Innovative control architecture could be a solution helping the renewable energy microgrids flourish in terms of applicability and functionality
IP10	• The project embraces financial (a 10-year payback period considering energy cost reduction strategies), non-financial (social and environmental benefits) and technical advantages, including emergency power provision during a grid outage event • The microgrid's systems and equipment are aimed at providing flexibility with respect to market condition while ensuring microgrid sustainable operation • In this project, a battery and microgrid system are installed and implemented in the existing facilities, which can be a stimulus for the implementation of smart grid components and concepts, including the integration of control systems. This helps the operator improve energy efficiency and thus significantly reduce costs	• Keeping the technical vendors in touch is of the challenges that the microgrid implementation project face • Given the fact that the Internet of Energy featuring plug-and-play has not fully been developed, interoperation of sensors and controllers requires advanced interface programs and applications	• Maintenance agreements with main vendors with training on-site staff considered is required to prevent issues related to battery storage system installation • Because the power injected to the grid is subject to uncertainty, especially due to the fact that the facility is in need of battery for demand charge management, the bidding strategy for the wholesale market should be based on the extra battery capacity

Appendix C

Table C.1 National and international terminology standards

Type	Standard	Year	Title
National	AS 5711 [35]	2013	"Smart grid vocabulary"
International	IEC 60050 series	1983–2022	"International Electro-technical Vocabulary"

Table C.2 National and international electrical system operation standards

Type	Standard	Year	Title
National	AS 5577 [36]	2013	"Electricity network safety management systems"
	AS 5711 [35]	2013	"Smart grid vocabulary"
	AS 1284 series (Part 1:2004-Rec:2019 [37], Part 4:2006-Rec:2016 [38], Part 10.2:2006-Rec:2016 [39], Part 11:1995-Rec:2019 [40], Part 12:1995-Rec:2019 [41], Part 13:2002-Rec:2016 [42])	1973–2019	"Electricity metering"
	AS 62052 series (AS 62052.11:2018 [43], AS 62052.21:2018 [44], AS 62052.31:2017-Amd1:2021 [45])	2005–2021	"Electricity metering equipment (AC)—general requirements"
	AS 62053 series	2003–2021	"Electricity metering equipment (AC)—particular requirements"

(continued)

I. A. Ibrahim et al., *Microgrids Design and Operation*, Power Systems,
https://doi.org/10.1007/978-981-96-4914-3

Table C.2 (continued)

Type	Standard	Year	Title
	AS 62054 series (AS 62054.11:2018 [46], AS 62054.21:2018 [47])	2006–2018	"Electricity metering equipment (AC)—tariff and load control"
	AS 62056.21 [48, 49]	2006 (Rec:2016)	"Electricity metering—data exchange for meter reading, tariff and load control—part 21: direct local data exchange"
International	IEC 61158 series	2007—2019	"Industrial communications networks—fieldbus specifications"
	IEC 61334 series	1995–2001	"Distribution automation using distribution line carrier systems"
	IEC 61850 series	2012–2022	"Communication networks and systems for power utility automation"
	IEC 61968 series	2011–2021	"Application integration at electric utilities—system interfaces for distribution management"
	IEC 61970 series	2005–2022	"Energy management system application program interface"
	IEC 62056 series	1998–2018	"Electricity metering data exchange"
	IEC 62325 series	2005–2022	"Framework for energy market communications"
	IEC 62351 series	2007–2022	"Power systems management and associated information exchange—data and communications security"
	IEC 62361 series (IEC 62361-2:2013 [50], IEC 62361-100:2016 [51], IEC 62361-102:2018 [52], IEC 62361-103:2018 [53])	2013–2018	"Power systems management and associated information exchange—interoperability in the long term"
	ISO 16484 series (Part 5:2017 [54])	2004–2020	"Building automation and control systems"
	ISO/IEC 14908 series (Part 1:2012: [55], Part 2:2012: [56], Part 3:2012: [57], Part 4:2012: [58])	2012	"Information technology—control network protocol"

Table C.3 National and international generation technologies standards

Type	Standard	Year	Title
National	AS/NZS 4777 series (AS/NZS 4777.1:2016 [59], AS/NZS 4777.1:2020-Amd1:2021 [60])	2005–2021	"Grid connection of energy systems via inverters"
	AS/NZS 4509 series (AS/NZS 4509.1:2009 [61, 62], AS/NZS 4509.2:2010-Rec:2017 [63, 64])	1999–2017	"Stand-alone power systems"
	AS/NZS 3010 [65, 66]	2017 (Amd:2020)	"Electrical installations—generating sets"
	AS/NZS 5033 [67]	2021	"Installation and safety requirements for photovoltaic (PV) arrays"
International	ISO 14084 series (Part 1:2015-Rec:2020 [68], Part 2:2015-Rec:2020 [69])	2015 (Rec:2020)	"Process diagrams for power plants"
	IEC 60255 series	1980–2021	"Measuring relays and protection equipment (Formerly known as Electrical relays)"
	IEC 61400 series	2012–2022	"Wind energy generation systems (Wind turbines)"
	IEC 60045-1 [70]	2020	"Steam turbines—part 1: specifications"
	IEC 60953 series (part 1:1990 [71], part 2:1990 [72], part 3:2001 [73])	1990–2001	"Rules for steam turbine thermal acceptance tests"
	IEC 62600 series	2012–2022	"Marine energy—wave, tidal and other water current converters"
	IEC 62270 [74]	2013	"Guide for computer-based control for hydroelectric power plant automation"
	IEC 62006 [75]	2010	"Hydraulic machines—acceptance tests of small hydroelectric installations"
	IEC 62862 series	2017–2022	"Solar thermal electric plants"
	IEC 60904 series	2011–2022	"Photovoltaic devices"
	IEC 62446-1 [76]	2016 (Amd1:2018)	"Photovoltaic (PV) systems—Requirements for testing, documentation and maintenance—part 1: grid connected systems—documentation, commissioning tests and inspection"

Table C.4 National and international asset management standards

Type	Standard	Year	Title
National	AS ISO 55000 [77]	2014	"Asset management—overview, principles and terminology"
	AS ISO 55001 [78]	2014	"Asset management—management systems—requirements"
	AS ISO 55002 [79]	2019	"Asset management—management systems—guidelines for the application of ISO 55001"
International	ISO 55000 [80]	2014	"Asset management—overview, principles and terminology"
	ISO 55001 [81]	2014	"Asset management—management systems—requirements"
	ISO 55002 [82]	2014	"Asset management—management systems—guidelines for the application of ISO 55001"

Table C.5 National and international system security standards

Type	Standard	Year	Title
National	HB 167 [83]	2006	"Security risk management"
	AS 1725 series	2010–2020	"Chain link fabric fencing—security fences and gates—general requirements"
	AS/NZS 3016 [84]	2003	"Electrical installations—electric security fences"
	AS/NZS 4421 [85]	2011	"Guard and patrol security services"
International	ISO 18788 [86]	2015	"Management system for private security operations—requirements with guidance for use"

Table C.6 National and international critical infrastructure resilience standards

Type	Standard	Year	Title
National	AS 3745 [87] (AS 3745:2010-Amd1:2014 [88], AS 3745:2010—Amd2:2018 [89])	2010 (Amd:2018)	"Planning for emergencies in facilities"
International	ISO 22301 [90]	2019	"Security and resilience—business continuity management systems—requirements"
	ISO 22313 [91]	2020	"Security and resilience—business continuity management systems—guidance on the use of ISO 22301"
	ISO/TS 22317 [92]	2021	"Security and resilience—business continuity management systems—guidelines for business impact analysis"
	ISO 22320 [93]	2018	"Security and resilience—emergency management—guidelines for incident management"
	ISO 28000 [94]	2022	"Security and resilience—security management systems—requirements"

Table C.7 National and international substations standards

Type	Standard	Year	Title
National	AS 62271 series	2005–2022	"High voltage switchgear and control gear"
	AS 2067 [95]	2016	"Substations and high voltage installations exceeding 1 KV ac"
International	IEC 61850 series	2012–2022	"Communication networks and systems for power utility automation"

Table C.8 National and international transformers standards

Type	Standard	Year	Title
National	AS/NZS 60076 series	2005–2020	"Power transformers"
	AS 61869 series	2021	"Instrument transformers"
International	IEC 60076 series	1997–2020	"Power transformers"
	IEC 61378 series (Part 1:2011 [96], Part 3:2015 [97])	2011–2015	"Converter transformers"
	IEC 62032 [98]	2012	"Guide for the application, specification and testing of phase-shifting transformers"

Table C.9 National and international switchgears standards

Type	Standard	Year	Title
National	AS 62271 series	2005–2022	"High voltage switchgear and controlgear"
International	IEC 62271 series	2009–2022	"High-voltage switchgear and controlgear"
	IEC TS 63058 [99]	2021	"Switchgear and controlgear and their assemblies for low voltage—environmental aspects"

Table C.10 National and international protection and relays standards

Type	Standard	Year	Title
National	–	–	–
International	IEC 61810 series	2006–2020	"Electromechanical elementary relays"
	IEC 62314 [100]	2006	"Solid-state relays"
	IEC 61812-1 [101]	2011	"Time relays for industrial and residential use"
	IEC 60255 series	1980–2021	"Measuring relays and protection equipment (Formerly known as Electrical relays)"

Table C.11 National and international cable and lines standards

Type	Standard	Year	Title
National	AS/NZS 7000 [102]	2016	"Overhead line design"
International	IEC 60227 series	1992–2011	"Polyvinyl chloride insulated cables of rated voltages up to and including 450/750 V"
	IEC 60502 (Part 1:2021 [103], Part 2:2014 [104], Part 4:2010 [105])		"Power cables with extruded insulation and their accessories for rated voltages from 1 kV (Um = 1.2 kV) up to 30 kV (Um = 36 kV)"
	IEC 60811 series	2007–2018	"Electric and optical fiber cables—test methods for non-metallic materials"
	IEC 60840 [106]	2020	"Power cables with extruded insulation and their accessories for rated voltages above 30 kV (Um= 36 kV) up to 150 kV (Um = 170 kV)—test methods and requirements"

Table C.12 National and international grid-side energy storages standards

Type	Standard	Year	Title
National	AS/NZS 4777 series (AS/NZS 4777.1:2016 [59], AS/NZS 4777.1:2020-Amd1:2021 [60])	2005–2021	"Grid connection of energy systems via inverters"
International	IEC 62282 series	2010–2021	"Fuel cell technologies"
	IEC 62933 series	2017–2022	"Electrical Energy Storage (EES) system"

Table C.13 National and international process automation systems standards

Type	Standard	Year	Title
National	AS 61508 series	1999–2015	"Functional safety of electrical/ electronic/programmable electronic safety-related systems"
	AS ISO 14649.1 [107, 108]	2004 (Rec:2015)	"Industrial automation systems and integration—physical device control—data model for computerised numerical controllers, part 1: overview and fundamental principles"
International	IEC 61508 series	2010	"Functional safety of electrical/ electronic/programmable electronic safety-related systems"
	ISO 14649 series	2003–2020	"Industrial automation systems and integration"

Table C.14 National and international electromagnetic standards

Type	Standard	Year	Title
National	AS/NZS 61000 series (Some parts are withdrawn and replaced by AS IEC 61000 series of standards)	1999–2021	"Electromagnetic compatibility (EMC)"
International	IEC 61000 series	1998–2022	"Electromagnetic compatibility (EMC)"

Table C.15 National and international power quality standards

Type	Standard	Year	Title
National	AS/NZS IEC 61000.3.12:2013 [109]	2013	"Electromagnetic compatibility (EMC), part 3.12: limits—limits for harmonic currents produced by equipment connected to public low-voltage systems with input current >16 A and ≤75 A per phase"
International	IEC 62586 series (Part 1:2017 [110], Part 2:2017-Amd1:2021 [111])	2017–2021	"Power quality measurement in power supply systems"

Table C.16 National and international market systems standards

Type	Standard	Year	Title
National	AS 60870 series	1995–2015	"Tele-control equipment and systems"
International	IEC 61850 series	2012–2022	"Communication networks and systems for power utility automation"
	IEC 61968 series	2011–2021	"Application integration at electric utilities—system interfaces for distribution management"
	IEC 61970 series	2005–2022	"Energy management system application program interface"
	IEC 62056 series	1998–2018	"Electricity metering data exchange"
	IEC 62325 series	2005–2022	"Framework for energy market communications"
	IEC 62351 series	2007–2022	"Power systems management and associated information exchange—data and communications security"
	IEC 62361 series (IEC 62361-2:2013 [50], IEC 62361-100:2016 [51], IEC 62361-102:2018 [52], IEC 62361-103:2018 [53])	2013–2018	"Power systems management and associated information exchange—interoperability in the long term"

Table C.17 National and international cyber security standards

Type	Standard	Year	Title
National	AS ISO/IEC 27001 [112, 113]	2015 (Amd:2016)	"Information technology—security techniques—information security management systems—requirements"
International	ISO/IEC 27000 series	N/A	"Information technology—security techniques—information security management systems"
	IEC 62351 series	2007–2022	"Power systems management and associated information exchange—data and communications security"

Table C.18 National and international local energy storages standards

Type	Standard	Year	Title
National	AS/NZS 4777 series (AS/NZS 4777.1:2016 [59], AS/NZS 4777.1:2020-Amd1:2021 [60])	2005–2021	"Grid connection of energy systems via inverters"
	AS/NZS 4755.3.5 [114]	2016	"Demand response capabilities and supporting technologies for electrical products, part 3.5: interaction of demand response enabling devices and electrical products—operational instructions and connections for grid connected electrical energy storage (EES) systems"
	AS/NZS 5139 [115]	2019	"Electrical installations—safety of battery systems for use with power conversion equipment"
International	IEC 62933 series	2017–2022	"Electrical Energy Storage (EES) system"

Table C.19 National and international distributed energy coordination standards

Type	Standard	Year	Title
National	–	–	–
International	IEC TS 62786 [116]	2017	"Distributed energy resources connection with the grid"

Table C.20 National and international BMS standards

Type	Standard	Year	Title
National	AS/NZS 11801.1 [117]	2019	"Information technology—Generic cabling for customer premises part 1: general requirements (ISO/IEC 11801-1:2017, MOD)"
	AS/NZS 14763 series	2007–2022	"Information technology—implementation and operation of customer premises cabling"
International	ISO 16484 series (Part 5:2017 [54])	2004–2020	"Building automation and control systems"
	IEC 60870 series	1988–2022	"Tele-control equipment and systems"
	IEC 60904 series	2011–2022	"Photovoltaic devices"
	IEC 61158 series	2007–2019	"Industrial communications networks—fieldbus specifications"
	IEC 62056 series	1998–2018	"Electricity metering data exchange"
	IEC 62351 series	2007–2022	"Power systems management and associated information exchange—data and communications security"
	ISO/IEC 14543 series	2000–2020	"Information technology—home electronic system architecture"
	ISO 16484 series (Part 5:2017 [54])	2004–2020	"Building automation and control systems"
	ISO/IEC 14908 series (Part 1:2012: [55], Part 2:2012: [56], Part 3:2012: [57], Part 4:2012: [58])	2012	"Information technology—control network protocol"

Table C.21 National and international customer energy management standards

Type	Standard	Year	Title
National	–	–	–
International	IEC 62746 series (IEC 62746-10-1:2018 [118], IEC 62746-10-3:2018 [119], IEC TR 62746-2:2015 [120], IEC TS 62746-3:2015 [121])	2015–2018	"Systems interface between customer energy management system and the power management system"

Table C.22 National and international DRM equipment standards

Type	Standard	Year	Title
National	AS/NZS 4755 series (Part 1:2017 [122], Part 3.1:2014 [123], Part 3.2:2014 [124], Part 3.3:2014 [125], Part 3.5:2016 [114])	2007–2017	"Demand response capabilities and supporting technologies for electrical products"
International	ISO/IEC 15067-3 [126]	2012 (Rec:2017)	"Information technology—Home Electronic System (HES) application model—part 3: model of a demand-response energy management system for HES"

Table C.23 National and international advanced metering standards

Type	Standard	Year	Title
National	AS 1284 series (Part 1:2004-Rec:2019 [37], Part 4:2006-Rec:2016 [38], Part 10.2:2006-Rec:2016 [39], Part 11:1995-Rec:2019 [40], Part 12:1995-Rec:2019 [41], Part 13:2002-Rec:2016 [42])	1973–2019	"Electricity metering"
	AS 62052 series (AS 62052.11:2018 [43], AS 62052.21:2018 [44], AS 62052.31:2017-Amd1:2021 [45])	2005–2021	"Electricity metering equipment (AC)—general requirements"
	AS 62053 series	2003–2021	"Electricity metering equipment (AC)—particular requirements"
	AS 62054 series (AS 62054.11:2018 [46], AS 62054.21:2018 [47])	2006–2018	"Electricity metering equipment (AC)—tariff and load control"
	AS 62056.21 [48, 49]	2006 (Rec:2016)	"Electricity metering—Data exchange for meter reading, tariff and load control—part 21: direct local data exchange"

(continued)

Table C.23 (continued)

Type	Standard	Year	Title
International	IEC 60870 series	1988–2022	"Tele-control equipment and systems"
	IEC 60904 series	2011–2022	"Photovoltaic devices"
	IEC 61158 series	2007–2019	"Industrial communications networks—fieldbus specifications"
	IEC 61334 series	1995–2001	"Distribution automation using distribution line carrier systems"
	IEC 62056 series	1998–2018	"Electricity metering data exchange"
	IEC 62325 series	2005–2022	"Framework for energy market communications"
	IEC 62351 series	2007–2022	"Power systems management and associated information exchange—data and communications security"
	IEC 62443 series	2009–2020	"Industrial communication networks—network and system security"
	IEC 62541 series	2015–2020	"OPC unified architecture"

Table C.24 National and international EVSE standards

Type	Standard	Year	Title
National	AS 5732 [127]	2015	"Electric vehicle operation: electric vehicle maintenance and repair"
	AS ISO 6469 (Part 1:2014 [128], Part 2:2014 [129], Part 3:2014 [130])	2014	"Electrically propelled road vehicles—safety specifications"
	AS IEC 61851 series (AS IEC 61851.23:2014 [131], AS IEC 61851.24:2014 [132])	2014	"Electric vehicle conductive charging system"
	SA TR IEC 60783 [133]	2014	"Wiring and connectors for electric road vehicles"
	SA TR ISO 8713 [134]	2014	"Electrically propelled road vehicles—vocabulary"

(continued)

Table C.24 (continued)

Type	Standard	Year	Title
International	ISO 6469 series (ISO 6469.1:2019 [135], ISO 6469.4:2015-Rec:2021 [136])	2015–2019	"Electrically propelled road vehicles—safety specifications"
	ISO TR 8713 [137]	2019	"Electrically propelled road vehicles—vocabulary"
	ISO 8714 [138]	2002	"Electric road vehicles—reference energy consumption and range—test procedures for passenger cars and light commercial vehicles"
	ISO 8715 [139]	2001 (Rec:2017)	"Electric road vehicles—road operating characteristics"
	ISO 12405.4 [140]	2018	"Electrically propelled road vehicles—test specification for lithium-ion traction battery packs and systems—part 4: performance testing"
	ISO 17409 [141]	2020	"Electrically propelled road vehicles—conductive power transfer—safety requirements"

Table C.25 National and international inverters standards

Type	Standard	Year	Title
National	AS/NZS 4777 series (AS/NZS 4777.1:2016 [59], AS/NZS 4777.1:2020—Amd1:2021 [60])	2005–2021	"Grid connection of energy systems via inverters"
International	IEC 62109 (Part 1:2010 [142], Part 2:2011 [143], Part 3:2020 [144])	2010–2020	"Safety of power converters for use in photovoltaic power systems"

Table C.26 National and international communication standards

Type	Standard	Year	Title
National	AS 60870 series	1995–2015	"Tele-control equipment and systems"
International	IPv6	2017	"Internet Protocol version 6"
	ISO/IEC/IEEE 8802 series	2012–2021	"Information technology—telecommunications and information exchange between systems—local and metropolitan area networks"
	ISO/IEC TR 29181 series	2012–2017	"Information technology—future network—problem statement and requirements"
	IEC 61158 series	2007–2019	"Industrial communications networks—fieldbus specifications"
	IEC 61850 series	2012–2022	"Communication networks and systems for power utility automation"

Table C.27 National and international data standards

Type	Standard	Year	Title
National	AS 5711 [35]	2013	"Smart grid vocabulary"
International	IEC 62939 series (Part 1:2014 [145], Part 2:2018 [146])	2014–2018	"Smart grid user interface"
	ISO/IEC 29100 [147, 148]	2011 (Amd:2018)	"Information technology—security techniques—privacy framework"
	ISO/IEC 27018 [149]	2019	"Information technology—security techniques—code of practice for protection of personally identifiable information (PII) in public clouds acting as PII processors"

Table C.28 National and international digital twin standards

Type	Standard	Year	Title
National	–	–	–
International	IEC 62832 series (part 1:2020 [150], Part 2:2020 [151], Part 3:2020 [152])	2020	"Industrial-process measurement, control and automation—digital factory framework"
	IEEE P2806 [153] (P P2806.1:2020 [154])	2019–2020	"System architecture of digital representation for physical objects in factory environments"
	IEC PAS 63088 [155]	2017	"Smart manufacturing—reference architecture model industry 4.0 (RAMI4.0)"
	ISO/TS 18101-1 [156]	2019	"Automation systems and integration—oil and gas interoperability—part 1: overview and fundamental principles"
	ISO 23247 series	2021	"Automation systems and integration—digital twin framework for manufacturing"
	ISO/TR 24464 [157]	2020	"Automation systems and integration—industrial data—visualisation elements of digital twin"

References

1. Bunker K, Doig S, Hawley K, Morris J (2015) Renewable microgrids: profiles from islands and remote communities across the globe
2. ABB (2014) Case study for coral bay energy modelling for a wind/diesel and stabilization hybrid system
3. Hydro Tasmania (2010) King island renewable energy integration project. https://www.hydro.com.au/clean-energy/hybrid-energy-solutions/success-stories/king-island. Accessed 13 May 2022
4. Brookfield Energy Australia Pty Limited (2017) Delivering higher renewable penetration in new land & housing developments through edge-of-grid microgrids. Hunter Valley, New South Wales
5. Moreland Energy Foundation Ltd (2017) Here comes the sun. Melbourne, Victoria
6. Drager P (2017) DeGrussa solar project. Meekatharra, WA, Australia
7. Dickeson G, McLeod L, Dobb A et al (2019) Ramp rate control for Pv plant integration: experience from Karratha airport's hybrid power station. In: 36th European photovoltaic solar energy conference and exhibition
8. Harris R (2018) Karratha airport solar project. Karratha Airport, WA, Australia
9. LO3 Energy Pty Ltd (2019) Latrobe valley microgrid feasibility assessment. Latrobe Valley, Victoria
10. ACTO (2019) Presentation: ATCO clean energy innovation hub lessons
11. ATCO (2019) Clean energy innovation hub. Jandakot, Western Australia
12. Horizon Power (2019) Onslow distributed energy resources management system (DERMS). https://www.horizonpower.com.au/your-community/getting-future-ready/onslow-distributed-energy-resources-management-system-derms/. Accessed 18 May 2022
13. Indra Australia Pty Limited (2019) Smart energy city introductory report. Clayton, VIC
14. Carnegie Clean Energy Ltd. (2020) Garden island microgrid report. Garden Island, Western Australia
15. Moghbel M, Calais M, Glenister S et al (2020) Lessons learnt report # 1—Carnarvon distributed energy resources (DER) trials. Western Australia, Australia
16. Moghbel M, Calais M, Glenister S et al (2020) Horizon power Carnarvon distributed energy resources (DER) trials. Western Australia, Australia
17. Western Power (2021) Study into the feasibility of a microgrid at Kalbarri. Kalbarri, WA, Australia

I. A. Ibrahim et al., *Microgrids Design and Operation*, Power Systems, https://doi.org/10.1007/978-981-96-4914-3

18. Synergy (2021) Alkimos beach energy storage trial
19. Ford H, Firestone H, Carrel A et al (2021) Yackandandah microgrid development trial. Yackandandah, VIC
20. Totally Renewable Yackandandah, Mach 2 Consulting (2021) Yackandandah microgrid feasibility study. Yackandandah, VIC
21. Loveday J, Morrison GM, Hansen P et al (2021) Increasing the uptake of solar PV and battery storage in strata residential developments. White Gum Valley, City of Fremantle, WA
22. PGWF Pty Ltd (2021) Advanced energy resources lessons learnt
23. Lord Howe Island Board (LHIB) (2021) Lord Howe island renewable energy project system design report. Lord Howe Island, New South Wales
24. Hydro Tasmania (2018) Flinders island hybrid energy hub. Flinders Island, Tasmania
25. Power H (2020) Horizon power Denham hydrogen demonstration. https://arena.gov.au/projects/horizon-power-denham-hydrogen-demonstration/. Accessed 19 May 2022
26. Sifnos Island Cooperative (2016) Energy autonomy of Sifnos. https://sifnosislandcoop.gr/en/energyautonomy/index.html#project. Accessed 27 May 2022
27. Katsaprakakis DA, Dakanali E, Dimopoulos A, Gyllis Y (2022) Energy transition on Sifnos: an approach to economic and social transition and development. Appl Sci 12. https://doi.org/10.3390/app12052680
28. Santarelli M, Torino P (2021) Remote area energy supply with multiple options for integrated hydrogen-based technologies
29. Number WP, Marocco P, Ferrero D et al (2018) Remote area energy supply with multiple options for integrated hydrogen-based technologies
30. Cleveland O, Ali Ahmed Andrew R Thomas OH, Henning M (2018) Techno-economic feasibility analysis of a microgrid in downtown Cleveland, Ohio
31. Schatz Energy Research Center, California Energy Commission. Energy Research and Development Division (2019) Demonstrating a secure, reliable, low-carbon community microgrid at the Blue Lake Rancheria
32. Katmale H, Clark S, Bialek TO et al (2019) Borrego Springs: California's first renewable energy-based community microgrid
33. Bliss D (2019) A novel, renewable energy microgrid for a California Healthcare Facility California Energy Commission
34. Positas L, Microgrid C, Patterson J (2019) Making a microgrid from legacy systems
35. Standards Australia (2013) Smart grid vocabulary. AS 5711-2013 1–27
36. Standards Australia (2013) Electricity network safety management systems. AS 5577-2013 1–17
37. Standards Australia (2019) Electricity metering, part 1: general purpose induction watthour meters. AS 1284.1-2004 REC2019
38. Standards Australia (2016) Electricity metering—part 4: socket mounting system. AS 1284.4-2006 Rec2016
39. Standards Australia (2016) Electricity metering—part 10.2: data exchange for meter reading, tariff and load control—direct local data exchange via hand-held unit (HHU)—ANSI Standard interface. AS 1284.10.2-2006 Rec2016
40. Standards Australia (2019) Electricity metering, part 11: single-phase multifunction watthour meters. AS 1284.11-1995 REC2019
41. Standards Australia (2019) Electricity metering, part 12: polyphase multifunction (non-demand) watthour meters (Class 1). AS 1284.12-1995 REC2019
42. Standards Australia (2016) Electricity metering—part 13: in-service compliance testing. AS/NZS 1284.13:2002 Rec:2016
43. Standards Australia (2018) Electricity metering equipment (ac)—general requirements, tests and test conditions metering equipment (IEC 62052-11:2016 (ED.1.1) MOD). AS 62052.112018 1–55
44. Standards Australia (2018) Electricity metering equipment (ac)—general requirements, tests and test conditions, part 21: tariff and load control equipment (IEC 62052-21:2016 (ED.1.1) MOD). AS 62052.212018

45. Standards Australia (2021) Electricity metering equipment (AC)—general requirements, tests and test conditions, part 31: product safety requirements and tests (IEC 62052-31:2015 (ED.1.0) MOD). AS 62052.312017 Amd 12021
46. Standards Australia (2018) Electricity metering equipment (a.c.)—tariff and load control, part 11: particular requirements for electronic ripple control receivers (IEC 62054-11:2004 (ED.1.1), MOD). AS 62054.112018
47. Standards Australia (2018) Electricity metering (AC)—tariff and load control, part 21: particular requirements for time switches (IEC 62054-21:2017 (ED.1.1), MOD). AS 62054.212018
48. Standards Australia (2006) Electricity metering—data exchange for meter reading, tariff and load control—direct local data exchange. AS 62056.21-2006
49. Standards Australia (2016) Electricity metering—data exchange for meter reading, tariff and load control—part 21: direct local data exchange. AS 62056.21-2006 Rec2016
50. IEC (2013) Power systems management and associated information exchange—interoperability in the long term—part 2: end to end quality codes for supervisory control and data acquisition (SCADA). IEC 62361-22013 1–101
51. IEC (2016) Power systems management and associated information exchange—interoperability in the long term—part 100: CIM profiles to XML schema mapping. IEC 62361-1002016 1–105
52. IEC (2018) Power systems management and associated information exchange—interoperability in the long term—part 102: CIM—IEC 61850 harmonization. IEC TS 62361-1022018 1–119
53. IEC (2018) Power systems management and associated information exchange—interoperability in the long term—part 103: standard profiling. IEC TR 62361-1032018 1–26
54. ISO (2017) Building automation and control systems (BACS)—part 5: data communication protocol. ISO 16484-52017 1–1327
55. ISO/IEC (2012) Information technology—control network protocol—part 1: protocol stack. ISO/IEC 14908-1:2012 1–420
56. ISO/IEC (2012) Information technology—control network protocol—part 2: twisted pair communication. ISO/IEC 14908-2:2012 1–16
57. ISO/IEC (2012) Information technology—control network protocol—part 3: power line channel specification. ISO/IEC 14908-3:2012 1–13
58. Standards Australia (2016) Grid connection of energy systems via inverters—part 1: installation requirements. AS/NZS 4777.1:2016
59. Standards Australia (2015) AS/NZS 4777.2:2015: grid connection of energy systems via inverters—part 2: inverter requirements
60. ISO/IEC (2012) Information technology—control network protocol—part 4: IP communication. ISO/IEC 14908-4:2012 1–52
61. Standards Australia (2009) Stand-alone power systems, part 1: safety and installation. AS/NZS 4509.1:2009
62. Standards Australia (2017) Stand-alone power systems—part 1: safety and installation. AS/NZS 4509.1:2009 Rec:2017
63. Standards Australia (2010) Stand-alone power systems, part 2: system design. AS/NZS 4509.2:2010
64. Standards Australia (2017) Stand-alone power systems—part 2: System Design. AS/NZS 4509.2:2010 Rec:2017
65. Standards Australia (2017) Electrical installations—generating sets. AS/NZS 3010:2017
66. Standards Australia (2020) Electrical installations—generating sets. AS/NZS 30102017 Amd 12020
67. Standards Australia (2021) Installation and safety requirements for photovoltaic (PV) arrays. AS/NZS 5033:2021
68. ISO (2015) Process diagrams for power plants—part 1: specification for diagrams. ISO 14084-12015 1–28
69. ISO (2015) Process diagrams for power plants—part 2: graphical symbols. ISO 14084-22015 1–60

70. IEC (2020) Steam turbines—part 1: specifications. IEC 60045-12020 1–146
71. IEC (1990) Rules for steam turbine thermal acceptance tests. Part 1: method A—high accuracy for large condensing steam turbines. IEC 60953-11990 1–163
72. IEC (1990) Rules for steam turbine thermal acceptance tests. Part 2: method B—wide range of accuracy for various types and sizes of turbines. IEC 60953-21990 1–233
73. IEC (2001) Rules for steam turbine thermal acceptance tests—part 3: thermal performance verification tests of retrofitted steam turbines. IEC 60953-32001 1–157
74. IEC (2013) Guide for computer-based control for hydroelectric power plant automation. IEC 622702013 1–148
75. IEC (2010) Hydraulic machines—acceptance tests of small hydroelectric installations. IEC 620062010 1–223
76. IEC (2018) Photovoltaic (PV) systems—requirements for testing, documentation and maintenance—part 1: grid connected systems—documentation, commissioning tests and inspection. IEC 62446-12016+AMD12018 Consol. Version 1–171
77. Standards Australia (2014) Asset management - Overview, principles and terminology. AS ISO 550002014
78. Standards Australia (2014) Asset management—management systems—requirements. AS ISO 550012014
79. Standards Australia (2019) Asset management—management systems—guidelines for the application of ISO 55001. AS ISO 550022019
80. ISO (2010) ISO 55000:2014 draft. Asset management—overview, principles and terminology. ISO 550002014 23
81. ISO (2014) Asset management—management systems—requirements. ISO 550012014 1–14
82. ISO (2018) Asset management—management systems—guidelines for the application of ISO 55001. ISO 550022018 1–72
83. Standards Australia (2006) Security risk management. HB 1672006
84. Standards Australia (2003) Electrical installations—electric security fences. AS/NZS 3016:2002 1–16
85. Standards Australia (2011) Guard and patrol security services. AS/NZS 4421:2011 1–32
86. ISO (2015) Management system for private security operations—requirements with guidance for use. ISO 187882015 1–98
87. Standards Australia (2010) Planning for emergencies in facilities. AS 3745-2010
88. Standards Australia (2014) Planning for emergencies in facilities. AS 3745-2010/Amdt 12014
89. Standards Australia (2018) Planning for emergencies in facilities. AS 3745-2010 Amd 22018
90. ISO (2019) Security and resilience—business continuity management systems—requirements. ISO 223012019 1–21
91. ISO (2020) Security and resilience—business continuity management systems—guidance on the use of ISO 22301. ISO 223132020
92. ISO (2021) Security and resilience—business continuity management systems—guidelines for business impact analysis. ISO/TS 22317:2021 1–36
93. ISO (2018) Security and resilience—emergency management—guidelines for incident management. ISO 223202018 1–20
94. ISO (2022) Security and resilience—security management systems—requirements. ISO 280002022(en)
95. Standards Australia (2016) Substations and high voltage installations exceeding 1 kV a.c. AS 20672016
96. IEC (2011) Converter transformers—part 1: transformers for industrial applications. IEC 61378-12011 1–184
97. IEC (2015) Converter transformers—part 3: application guide. IEC 61378-32015 1–192
98. IEC (2012) Guide for the application, specification and testing of phase-shifting transformers. IEC 620322012 1–42
99. IEC (2021) Switchgear and controlgear and their assemblies for low voltage—environmental aspects. IEC TS 630582021 1–112
100. IEC (2006) Solid-state relays. IEC 623142006 1–38

101. IEC (2011) Time relays for industrial and residential use—part 1: requirements and tests. IEC 61812-12011 1–89
102. Standards Australia (2016) Overhead line design. AS/NZS 7000:2016 1–304
103. IEC (2021) Power cables with extruded insulation and their accessories for rated voltages from 1 kV up to 30 kV—part 1: cables for rated voltages of 1 kV and 3 kV. IEC 60502-12021 Redline version 1–116
104. IEC (2014) Power cables with extruded insulation and their accessories for rated voltages from 1 kV up to 30 kV—part 2: cables for rated voltages from 6 kV up to 30 kV. IEC 60502-22014 Redline version 1–265
105. IEC (2010) Power cables with extruded insulation and their accessories for rated voltages from 1 kV up to 30 kV—part 4: test requirements on accessories for cables with rated voltages from 6 kV up to 30 kV. IEC 60502-42010 1–62
106. IEC (2020) Power cables with extruded insulation and their accessories for rated voltages above 30 kV up to 150 kV—test methods and requirements. IEC 608402020 Redline version 1–264
107. Standards Australia (2004) Industrial automation systems and integration—physical device control—data model for computerized numerical controllers, part 1: overview and fundamental principles. AS ISO 14649.1-2004
108. Standards Australia (2015) Industrial automation systems and integration—physical device control—data model for computerized numerical controllers—part 1: overview and fundamental principles. AS ISO 14649.1-2004 Rec2015
109. Standards Australia (2013) Electromagnetic compatibility (EMC), part 3.12: limits—limits for harmonic currents produced by equipment connected to public low-voltage systems with input current >16 A and ≤75 A per phase. AS/NZS IEC 61000.3.122013
110. IEC (2017) Power quality measurement in power supply systems—part 1: power quality instruments (PQI). IEC 62586-12017 1–71
111. IEC (2021) Power quality measurement in power supply systems—part 2: functional tests and uncertainty requirements. IEC 62586-22017+AMD12021 Consol. version 1–617
112. Standards Australia (2015) Information technology—security techniques—information security management systems—requirements. AS ISO/IEC 270012015
113. Standards Australia (2016) Information technology—security techniques—information security management systems—requirements. AS ISO/IEC 270012015 Amd 12016
114. Standards Australia (2016) Demand response capabilities and supporting technologies for electrical products, part 3.5: interaction of demand response enabling devices and electrical products—operational instructions and connections for grid-connected electrical energy storage (EE). AS/NZS 4755.3.5:2016 1–76
115. Standards Australia (2019) Electrical installations—safety of battery systems for use with power conversion equipment. AS/NZS 5139:2019 1–172
116. IEC (2017) Distributed energy resources connection with the grid. IEC TS 627862017 1–25
117. Standards Australia (2019) Information technology—generic cabling for customer premises part 1: general requirements (ISO/IEC 11801-1:2017, MOD). AS/NZS 11801.1:2019 1–181
118. IEC (2018) Systems interface between customer energy management system and the power management system—part 10-1: open automated demand response. IEC 62746-10-12018 1–207
119. IEC (2018) Systems interface between customer energy management system and the power management system—part 10-3: open automated demand response—adapting smart grid user interfaces to the IEC common information model. IEC 62746-10-32018 1–58
120. IEC (2015) Systems interface between customer energy management system and the power management system—part 2: use cases and requirements. IEC TR 62746-22015 1–350
121. IEC (2015) Systems interface between customer energy management system and the power management system—part 3: architecture. IEC TS 62746-32015 1–35
122. Standards Australia (2017) Demand response capabilities and supporting technologies for electrical products. AS/NZS 4755.1:2017 1–36

123. Standards Australia (2014) Demand response capabilities and supporting technologies for electrical products, part 3.1: interaction of demand response enabling devices and electrical products—operational instructions and connections for air conditioners. AS/NZS 4755.3.1:2014
124. Standards Australia (2014) Demand response capabilities and supporting technologies for electrical products—part 3.2: interaction of demand response enabling devices and electrical products—operational instructions and connections for devices controlling swimming pool pump-un. AS/NZS 4755.3.2:2014 1–40
125. Standards Australia (2014) Demand response capabilities and supporting technologies for electrical products, part 3.3: interaction of demand response enabling devices and electrical products—operational instructions and connections for electric storage and electric-boosted storage. AS/NZS 4755.3.3:2014
126. ISO/IEC (2012) Information technology—home electronic system (HES) application model—part 3: model of a demand-response energy management system for HES. ISO/IEC 15067-3:2012 1–34
127. Standards Australia (2015) Electric vehicle operations—maintenance and repair. AS 57322015
128. Standards Australia (2014) Electrically propelled road vehicles—safety specifications, part 1: on-board rechargeable energy storage system (RESS). AS ISO 6469.12014
129. Standards Australia Electrically propelled road vehicles—safety specifications, part 2: vehicle operational safety means and protection against failures. AS ISO 6469.22014
130. Standards Australia (2014) Electrically propelled road vehicles—safety specifications, part 3: protection of persons against electric shock. AS ISO 6469.32014
131. Standards Australia (2014) Electric vehicle conductive charging system, part 23: D.C electric vehicle charging station. AS IEC 61851.232014
132. Standards Australia (2014) Electric vehicle conductive charging system, part 24: digital communication between a d.c. EV charging station and an electric vehicle for control of d.c. charging. AS IEC 61851.242014
133. Standards Australia (2014) Wiring and connectors for electric road vehicles. SA TR IEC 607832014
134. Standards Australia (2014) Electrically propelled road vehicles—vocabulary. SA TR ISO 87132014
135. ISO (2019) Electrically propelled road vehicles—safety specifications—part 1: rechargeable energy storage system (RESS). ISO 6469-12019 1–26
136. ISO (2015) Electrically propelled road vehicles—safety specifications—part 4: post crash electrical safety. ISO 6469-42015 1–15
137. ISO (2019) Electrically propelled road vehicles—vocabulary. ISO/TR 8713:2019 1–38
138. ISO (2002) Electric road vehicles—reference energy consumption and range—test procedures for passenger cars and light commercial vehicles. ISO 87142002 1–25
139. ISO (2001) Electric road vehicles—road operating characteristics. ISO 87152001 1–17
140. ISO (2018) Electrically propelled road vehicles—test specification for lithium-ion traction battery packs and systems—part 4: performance testing. ISO 12405-42018 1–72
141. ISO (2020) Electrically propelled road vehicles—conductive power transfer—safety requirements. ISO 174092020 1–41
142. IEC (2010) Safety of power converters for use in photovoltaic power systems—part 1: general requirements. IEC 62109-12010 1–316
143. IEC (2011) Safety of power converters for use in photovoltaic power systems—part 2: particular requirements for inverters. IEC 62109-22011 1–64
144. IEC (2020) Safety of power converters for use in photovoltaic power systems—part 3: particular requirements for electronic devices in combination with photovoltaic elements. IEC 62109-32020 1–60
145. IEC (2014) Smart grid user interface—part 1: interface overview and country perspectives. IEC TR 62939-12014 1–128

146. IEC (2018) Smart grid user interface—part 2: architecture and requirements. IEC TS 62939-22018 1–15
147. ISO/IEC (2011) Information technology—security techniques—privacy framework. ISO/IEC 29100:2011 1–21
148. ISO/IEC (2018) Information technology—security techniques—privacy framework—amendment 1: clarifications. ISO/IEC 29100:2011/AMD 1:2018 1–4
149. ISO/IEC (2019) Information technology—security techniques—code of practice for protection of personally identifiable information (PII) in public clouds acting as PII processors. ISO/IEC 27018:2019 1–23
150. IEC (2020) Industrial-process measurement, control and automation—digital factory framework—part 1: general principles. IEC 62832-12020 1–66
151. IEC (2020) Industrial-process measurement, control and automation—digital factory framework—part 2: model elements. IEC 62832-22020 1–149
152. IEC (2020) Industrial-process measurement, control and automation—digital factory framework—part 3: application of digital factory for life cycle management of production systems. IEC 62832-32020 1–55
153. IEEE Computer Society (2019) System architecture of digital representation for physical objects in factory environments. IEEE P2806
154. IEEE Computer Society (2020) Standard for connectivity requirements of digital representation for physical objects in factory environments. IEEE P2806.1
155. IEC (2017) Smart manufacturing—reference architecture model industry 4.0 (RAMI4.0). IEC PAS 630882017 1–39
156. ISO (2019) Automation systems and integration—oil and gas interoperability—part 1: overview and fundamental principles. ISO/TS 18101-1:2019 1–26
157. ISO (2020) Automation systems and integration—industrial data—visualization elements of digital twins. ISO/TR 24464:2020 1–19

Uncited Reference

158. Anvari-Moghaddam A, Abdi H, Mohammadi-Ivatloo B, Hatziargyriou N (2021) Microgrids: advances in operation, control, and protection. Springer Nature

MIX
Papier aus verantwortungsvollen Quellen
Paper from responsible sources
FSC® C105338

FSC
www.fsc.org

If you have any concerns about our products,
you can contact us on
ProductSafety@springernature.com

In case Publisher is established outside the EU,
the EU authorized representative is:
Springer Nature Customer Service Center GmbH
Europaplatz 3, 69115 Heidelberg, Germany

Printed by Libri Plureos GmbH
in Hamburg, Germany